Patrick Moore
Practical Ast

D1579521

Springer

London
Berlin
Heidelberg
New York
Hong Kong
Milan
Paris
Tokyo

Other titles in this series

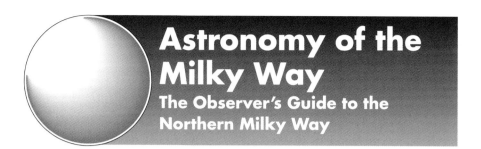

Astronomy of the Milky Way
The Observer's Guide to the Northern Milky Way

Mike Inglis

With 288 Figures
(including 20 in color)

Springer

Dr Michael D. Inglis, FRAS
State University of New York, USA

British Library Cataloguing in Publication Data
Inglis, Mike, 1954–
   Astronomy of the Milky Way
   The observer's guide to the northern Milky Way. – (Patrick
   Moore's practical astronomy series)
   1. Milky Way
   I. Title
   523.1′13
ISBN 1852337095

Library of Congress Cataloging-in-Publication Data
Inglis , Mike, 1954–
   Astronomy of the Milky Way/Mike Inglis.
      p. cm. – (Patrick Moore's practical astronomy series)
   Includes bibliographical references.
   Contents: pt. 1. The observer's guide to the northern Milky Way –
   pt. 2. The observer's guide to the southern Milky Way.
      ISBN 1–85233–709–5 (alk. paper) – ISBN 1–85233–709–5 (pt. 1) –
   ISBN 1–85233–742–7 (pt. 2)
      1. Milky Way. I. Title. II. Series.
   QB981.I45 2003
   523.1′13–dc21   2003050548

Patrick Moore's Practical Astronomy Series ISSN 1617–7185
ISBN 1–85233–709–5 Springer-Verlag London Berlin Heidelberg
a member of BertelsmannSpringer Science+Business Media GmbH
http://www.springer.co.uk

Typeset by EXPO Holdings, Malaysia
58/3830-543210   Printed on acid-free paper   SPIN 10861212

Dedicated to
the Astronomers of the Bayfordbury Observatory,
University of Hertfordshire,
past, present and future

Preface

Sometime during the last century when I was a boy, I remember looking at the night sky and being amazed at how bright and spectacular the Milky Way appeared as it passed through the constellation of Cygnus. It was an August Bank Holiday in the UK, and so, naturally, it was cold and clear. I may have looked at the Milky Way several times before that momentous evening but for some reason it seemed to stand out in a way it never had before. It was then that I began to observe the Milky Way as an astronomical entity in its own right and not as just a collection of constellations. It was also about that time that I had an idea for a book devoted to observing the Milky Way.

Fast forward a few years to a fortuitous meeting with John Watson, the astronomy editor of Springer-Verlag, who listened to my idea about a Milky Way book, and agreed that it would be a good idea. So I began, writing down notes and traveling the world, but at the same time observing hitherto uncharted regions of the sky (for me anyway!) and delighting in the new wonders it presented. After what seemed like an age, the book was completed, and you hold the finished product in your hands.

However, along the route I have been helped and guided by many people, both astronomers and nonastronomers, and I want to take this opportunity to thank them for taking part in what was a long-cherished labour of love. Firstly, my publisher, John Watson, whom I mentioned above and who has overseen the project from initial idea to completed book, and without whose help this book would never have seen the light of day. His knowledge of publishing and its many aspects is impressive. Add to this the fact that he is also an amateur astronomer himself, and you have a potent combination.

I have also been fortunate to have the company and friendship of amateur and professional astronomers worldwide, who freely gave advice and observational anecdotes that have subsequently appeared in the book. Amateur astronomers are a great bunch of people and none more so than Michael Hurrell and Don Tinkler, fellow members of the South Bayfordbury Astronomical Society. Their companionship has been a godsend, especially when life and its many problems seemed to be solely concerned with preventing me from ever finishing the book. Thank you, chaps!

I have also had the good fortune to be associated with many fine professional astronomers, and so I would like to especially thank Bob Forrest of the University of Hertfordshire Observatory at Bayfordbury for teaching me most of what I know about observational astronomy. Bob's knowledge of the techniques and application of all things observational is truly impressive, and it has been an honor to be at his side many times when he has been observing. Furthermore, I must mention Chris Kitchin, Iain Nicolson, Alan McCall and Lou Marsh, also from Bayfordbury, for not only teaching me astronomy and astrophysics, but for instilling in me a passion to share this knowledge with the rest of the world! I am privileged to have them all as friends.

Several nonastronomy colleagues have also made my day-to-day life great fun, with many unexpected adventures and Jolly Boys' Outings, and so it is only right and proper that the guilty be named – Bill, Pete, Andy and Stuart.

However, astronomy and the writing of books is, shall we say, only a meteor-sized concern, when compared to the cosmological importance of one's family. Without their support and love – especially when I was writing a book – patience and understanding, I would never have completed the project. Firstly, I must thank my partner and companion Karen, as we whiz together through space on our journey towards the constellation Hercules. Her patient acknowledgment that astronomers are strange people and that sometimes astronomy is *the* most important subject in the universe has made my life a wonder. At times, when it seemed as if I would never finish the book, and the road ahead looked bleak and cloud-covered, she would come into the study with a cup of tea, a Hobnob and a few gentle words of encouragement, and suddenly all was well with the world. Thank you, *Cariad*. Then there is my brother Bob. He is a good friend and a great brother and has been – and amazingly, still is – supportive of all I have tried to achieve. Finally, I want to thank Mam, who has been with me all the way, even from before I saw the Milky Way in the garden in St Albans. She tells me that she always knew I would be an astronomer, and that it comes as no surprise to her to know that her son still spends a disproportionate amount of time standing outside in the cold and dark in the dead of winter and the middle of the night!

To all who have helped me become an astronomer and who make my life a lot of fun, many thanks, and don't forget the best is yet to come.

Dr Mike Inglis
St. Albans,
Hertfordshire, UK

and

Lawrenceville,
New Jersey, USA

August 2003

Arrangement of Book 1

During the course of writing this book, it became very apparent that it was going to be big – too big in fact to be totally contained within a single volume! This would have negated the premise of it being an observing guide that would and should be used at the telescope, not counting the cost of such a large book. After discussion with the publishers, it was decided to divide the book into two volumes: Book 1, which you now hold in your hands, would cover those constellations in the Milky Way that transit during the summer and autumn months, from July to December, and are thus best placed for observation in the northern hemisphere (but not exclusively); whereas the accompanying volume, Book 2, would cover those Milky Way constellations that transit during the winter and spring months, and so are best seen from the southern hemisphere (but again not exclusively).

However, experienced astronomers will know that a considerable number of Milky Way constellations residing in the northern part of the sky can be relatively easily seen from the southern hemisphere, and vice versa. In addition, many constellations that straddle the celestial equator can be seen by both northern and southern hemisphere observers, and from an observational point of view this simply means that quite a significant amount of overlap can occur. It is therefore possible for an observer living in, say, the UK to make use of a considerable portion of Book 2 (which deals with the southern sky) and likewise for an observer in, say, Australia, to find Book 1 (covering the northern sky) quite useful.

In order to minimize repetition of data, and so reduce book size and cost, the duplication of information has been avoided where possible. However, to ensure that each book is self-contained, the introductory chapters, as well as the appendices and object index, are in each book.

Acknowledgments

I would like to thank the following people and organizations for their help and permission to quote their work and for the use of the data and software they provided:

- Michael Hurrell and Donald Tinkler of the South Bayfordbury Astronomical Society, England, for use of their observing notes.
- Robert Forrest, of the University of Hertfordshire Observatory at Bayfordbury, for many helpful discussions and practical tutorials over several years, on all matters observational.
- Dr Jim Collett, of the University of Hertfordshire, UK, for information pertaining to the Milky Way.
- Dr Stuart Young, formerly of the University of Hertfordshire, UK, for many informative discussions relating to the Milky Way.
- The astronomers at Princeton University, USA, for many helpful discussions on the Milky Way.
- The European Space Organization, for permission to use the *Hipparcos* and *Tycho* catalogues.
- Gary Walker, of the American Association of Variable Star Observers, for information on the many types of variable star.
- Cheryl Gundy, of the Space Telescope Science Institute, Baltimore, USA, for supplying astrophysical data on many of the objects discussed.
- Dr Chris Packham, of the University of Florida, for information relating to galaxies, particularly active galaxies.
- The Smithsonian Astrophysical Observatory, for providing data on many of the stars and star clusters.
- Richard Dibon-Smith of Toronto, Canada, for allowing me to quote freely the data from his books, *STARLIST 2000.0* and *The Flamsteed Collection*, and for the use of several of his computer programs.
- The Secretariat of the International Astronomical Union for information pertaining to the Milky Way.
- The publishers of *The SKY Level IV* astronomical software, Colorado, USA, for permission to publish the star charts.

I would also like to take this opportunity to thank several amateur astronomers who indulged me in order to discuss matters concerning the layout of the book. They are a great bunch of dedicated people: Dave Eagle (UK), Peter Grego (UK), Phil Harrington (USA), John McAnally (USA), Paul Money (UK), James Mullaney (USA), Wolfgang Steinicke (Germany), Don Tinkler (UK) and John Watson (UK).

I must also make a special mention of the astrophotographers who were kind enough to let me publish their images. They have taken the art and science of astrophotography to new heights, yet still remain awed and humbled by the beauty of the objects whose images they capture. In a time when the actual process of photography and CCD imaging seems to assume a higher importance than the very objects being imaged, this gifted and talented group of people remind us all that astronomy is a beautiful, unique and wondrous subject, and we would all do well to keep this in mind. They are (in alphabetical order): Matt BenDaniel (USA), Mario Cogo (Italy), Bert Katzung (USA), Dr Jens Lüdeman and members of the IAS (Germany), Axel Mellinger (Germany), Thor Olson (USA), SBAS (UK), Harald Strauss and members of the AAS (Austria), Chuck Vaughn (USA) and the Students for the Exploration and Development of Space (SEDS).

In developing a book of this type, which presents a considerable amount of detail, it is nearly impossible to avoid errors. If any arise, I apologize for the oversight, and I would be more than happy to hear from you. Also, if you feel that I have omitted a star or object of particular interest, again, please feel free to contact me at: astrobooks@earthlink.net. I can't promise to reply to all e-mails, but I will certainly read them.

Contents

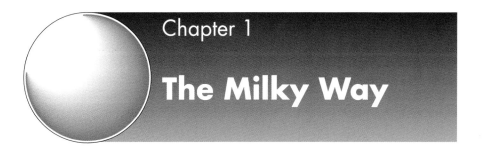

Chapter 1

The Milky Way

1.1 How to Use this Book

Most of us are familiar with the Milky Way. We may be lucky enough to live in a dark location and can see the misty band of light that stretches across the sky (see Figure 1.1). Others may live in an urban location and so can only glimpse the Milky way as a faint hazy patch that envelops several constellations. But how many of us make a point of observing the Milky Way as a celestial object in its own right? Very few I imagine, which is a pity as it holds a plethora of wonderful delights, ranging from deeply colored double and multiple star systems to immense glowing clouds of gas and mysterious dark nebulae which literally blacken the sky. It also holds quite a few star clusters that really do look like diamonds sprinkled on black velvet, not to mention the occasional neutron star, black hole and possible extra-solar planetary system! In fact, you could spend an entire career observing the Milky Way.

The Milky Way passes through many constellations; some are completely engulfed whilst others are barely brushed upon. Also it passes through both the northern and southern parts of the celestial sphere, making it a truly universal object, allowing it to be observed from anywhere in the world (see Star Chart 1).

The object of the book is not to give an introduction to the astrophysics of the Galaxy we live in, the Milky Way – there are many books listed in the appendices that can do that – but rather to introduce you to the many objects that can be observed that lie within the Milky Way. It may come as a surprise to you to know that you can observe the Milky Way on any clear night of the year, from any location on the Earth. So you could be observing from, say, Australia or Scotland. It wouldn't matter, as the Milky Way, or rather, particular parts of it, will be visible to you.

I have covered the complete Milky Way in this book, and so that means that there will be areas of it, and thus constellations, that may be unobservable from where you live. For instance, the constellation of Crux is a familiar one to observers living in Australia and New Zealand, but completely unobservable to European observers. Likewise, Camelopardalis and parts of Cepheus may be familiar friends to northern European observers, but are hidden from the view of our southern colleagues. What this does mean, however, that this is a truly universal book that can be used by any astronomer anywhere in the world.

Figure 1.1. The Milky Way (The Lünd Observatory).

Many of the objects mentioned will, of course, need some sort of optical equipment, but a significant number are naked-eye objects, which is appropriate as the Milky Way itself is a naked-eye object, and the biggest one at that! But there are also quite a considerable number that only require small telescopes or binoculars, and by small I mean, say, 6–10 cm aperture. There are also those objects that will require a somewhat larger aperture, say, 10–25 cm, and the majority of the faint objects are in this aperture range. But not to exclude those observers with large telescopes I have also mentioned a few objects where very large apertures will be needed. Thus there is something for all amateur astronomers to see.

But remember that the one pre-eminent type of object that is just perfect to view with binoculars, and is truly impressive, are the many rich and awe-inspiring star fields or star clouds. This is what makes the Milky Way so spectacular. On a clear evening, one can observe Cygnus, or Vulpecula, or Sagittarius, or Centaurus, and literally be transported to other realms. The sights that fill the field of view cannot really be described, and once seen are never forgotten.

It goes without saying that a good star atlas is an essential part of every amateur astronomer's arsenal and fortunately there are many fine atlases to be had. A fine example of an atlas that is perfect for naked-eye observing is the redoubtable *Norton's Star Atlas*. Armed with this, and perhaps a pair of binoculars, you will have a lifetime of opportunity. For those who need a more detailed atlas, there are two that warrant attention: *Sky Atlas 2000.0* and *Uranometria 2000*. Both of these cover most, if not all, of the objects mentioned in the book, and will allow you to locate and find most of the fainter and not easily recognizable objects. It is also possible these days to have planetarium software on a computer and these too are fine tools to have, many allowing detailed star charts to be printed.

An astute observer will notice that the boundary of the Milky Way that I have adopted may not be the same as that in, say, *Norton's*, or older star atlases. I have in fact taken the boundary to be that identified by the Dutch astronomer Antonie Pannakoek, who measured the density of stars in the sky, and ascribed a limiting factor to the star density which enabled a boundary to be placed on the visible Milky Way. We are of course completely immersed in the Milky Way and most stars, nebulae and clusters that we observe are in fact within the Milky Way.[1] Thus the misty band of light that we call the Milky Way is just a region of the sky where the number of stars is so large so as to make a distinct and visible impression.[2]

However, there is a downside to adopting this boundary, as many of the favorite objects are left out if they do not lie within the Milky Way. Examples of such passed-over showpieces are the Pleiades and the Andromeda Galaxy, to name just a couple. I actually had to be quite strict in this respect, as if I were to include those objects that just lie outside of the Milky Way, the size and cost of the book would have doubled.

The layout of the book is straightforward. I have grouped the constellations that lie within the Milky Way more or less[3] in order of the month at which they transit at midnight. This means that the constellation will be at its highest point above your horizon at midnight. The reason for this is quite simple: if I were to describe in detail all the Milky Way constellations that can be seen at any one particular time of the year, not only would I be repeating a substantial amount of information, but the book would probably be about 900 pages long! Thus, for January and February, I discuss the Milky Way in Monoceros and Canis Major, to name but a few. However, seasoned amateurs will know that there are other constellations besides these that are in the Milky Way that can be seen during these

---

[1] There are a few objects that can be observed that are actually located outside of the Milky Way (and I don't mean galaxies!).
[2] The boundary I use is also the one adopted by the International Astronomical Union.
[3] There are exceptions to this, as described in the text.

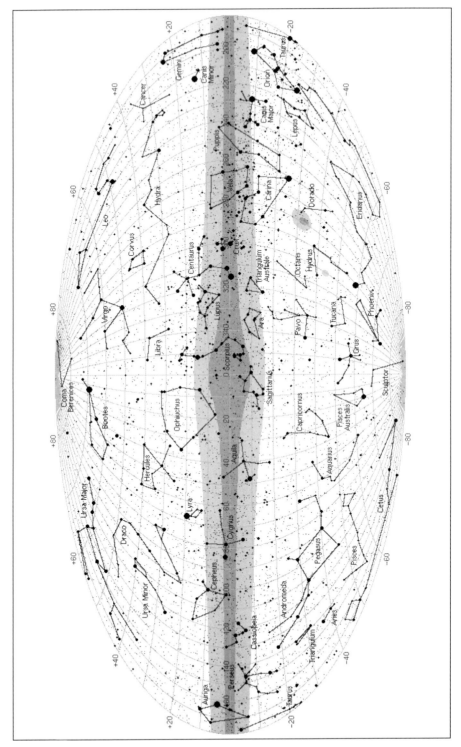

months, and this is perfectly true, but they do not transit at midnight during these months! The other constellations may be rising at midnight, or setting, or something in between, but they will not be at their best position for observation, *at midnight!* It is just a convenient means of presenting the data in a reasonable manner. You can of course view other parts of the sky during these months, say Cygnus, but it may not necessarily be at its optimum observing position. In fact, for the example given, it will be so low down as to be nearly unobservable. Nevertheless, it will be there for you to look at.[4]

For the sake of completeness, however, at the end of each chapter I have listed, for both northern and southern observers, those Milky Way constellations that are also visible but with the above caveat in mind! Armed with this knowledge, you can go out and observe quite a large proportion of the Milky Way on any clear night of the year, from anywhere in the world.

In addition, I do not structure each chapter in any formal way, but rather in a manner that seems appropriate. For instance, in Orion, I start off by describing the many wonderful double and multiple stars that the constellation has, whereas in Cassiopeia I begin with detailed descriptions of its many glorious star clusters.

Throughout the book are many simple star charts, and they are just that – simple! They are not meant to take the place of a star atlas, but are rather a pointer in the right direction. Also, I have had the opportunity to include quite a few wonderful photographs and CCD images of many of the objects described. These were taken by gifted and talented astrophotographers and astro-CCD imagers and to include them in the book is a privilege. You may notice, by their conspicuous absence, that there are no drawings of any of the objects in the book. The reason for this is simple. Not only can I not draw to save my life, but drawings or sketches, particularly of astronomical objects, are very personal constructs and more often than not do not resemble the generally recognized shape or form of an object. Rather, they describe what you, the observer, can see at that particular moment. It is no exaggeration to say that one can take two observers, show them the same object through two identical telescopes at the same time, and ask them to draw it, and you will end up with two quite different and distinct drawings. I believe it serves no useful purpose for me to include drawings of objects that show how I see a particular cluster or nebula, as it will be different from what you see. Furthermore, I agree with what the astronomer David Ratledge says in his book on the Caldwell objects, when he asks how can one really sketch something in detail when you are using averted vision?

Finally, at the end of each chapter is a list of the main objects mentioned in the text, giving their positions in right ascension and declination. This will allow you to use a star atlas, and the GOTO facility of your telescope or the setting circles on your telescope in order to locate and observe the objects. I also include, where appropriate, the objects' magnitude and, for double stars, their separation and position angles.

I have tried to include not only the well-known Messier, Caldwell and NGC objects that we are all familiar with, but also those that are perhaps less familiar to you. They may be faint and/or small, but they are all definitely worth observing. If I have left out an object

---

[4] I had quite a detailed correspondence with several amateur astronomers from the UK, Australia and the USA about how to present the data, and this method was the one that most of them preferred.

◄ **Star Chart 1.** The Milky Way in Galactic Coordinates. Compare this star map, that shows the Milky Way superimposed over the constellations, with the images of the Milky Way at the end of the book. Note that the contours of the Milky Way are approximate. Star map courtesy of Richard Powell.

that may be a particular favorite of yours, then I apologize, as I tried to include as much as I could.

So, enough of the words, but before we begin a year-round voyage of the Milky Way, I would like to take this opportunity to discuss a topic that is central to the subject matter of this book, and that every astronomer should be aware of.

1.2 A Plea to the Faithful

We live in a world where science and especially astronomy is making great leaps forward in our knowledge and understanding of the Universe. Every day there is a news article on some new discovery, either from an Earthbound telescope or satellite in space, or a new image is published of some magnificent and mysterious object deep in outer space.

At the same time more and more people are becoming interested in amateur astronomy. Telescopes are getting cheaper, better, and packed with additional extras like thousand-object databases and equatorial mounts. And, of course, the Internet is a vast resource of information on everything astronomical.

But one thing worries me – the ever increasing plague that is light pollution and especially when it concerns the topic of this book – The Milky Way. How many of us can remember a time when we could go out into our gardens or a nearby park and see the wonderful swathe of the Milky Way cut a path across the sky? Nowadays one needs to be deep in a sparsely populated rural landscape or high in the lonely mountains in order to see this wonder of nature.

We are told constantly that the resources and animals of the world we live in need to be conserved and protected, and I agree wholeheartedly with this notion. I have never seen a blue whale, or an American bison, or even a monarch butterfly or a slipper orchid. Furthermore, I have never visited the Great Barrier Reef or the Brazilian rainforest; yet I feel strongly that they must be protected for all and for ever and I am not a biologist or ecologist. In the same vein, we should keep the seas clean, the landscapes natural and the atmosphere breathable. Yet, in all this, it seems to me that the conservation of nature and the appreciation of our world stops when it gets dark. Surely, the most wonderful spectacle in all of nature is the night sky, blazing forth in all its glory. Yet most of the world seems unaware that we are losing this resource. In a recent study published in the *Monthly Notices of the Royal Astronomical Society*, it stated: "about one-fifth of the world population, more than two-thirds of the US population and more than one-half of the EU population have already lost naked-eye visibility of the Milky Way … and about two-thirds of the world population and 99% of the population in the US (excluding Alaska and Hawaii) and the EU live in areas where the night sky is above the threshold set for polluted status".[5] This is a truly appalling statistic! And if you think I am a zealot and maybe overly dramatic, just think back to your own experience in this regard. How much of the night sky have you seen literally disappear in just a few years?

I would like to think that in the future it would be possible for me to take my children or grandchildren out into the garden and show them the Milky Way, and how splendid it looks, hoping that it inspires the awe and wonder in them that it did and still does in me. But if we do not try to curb the energy- and resource-wasteful spread of light pollution, this will not happen. But what can we do?

---

[5] P. Cinzano, F. Falchi, C.D. Elvidge. *Monthly Notices of the Royal Astronomical Society*, **328**, 689–707 (2001).

In order to try to come to an equitable balance between conservation and common sense, we need to be aware of and appreciative to the wishes of the nonastronomer. It may be necessary to show them how beautiful, and, more importantly, how special the night sky and the Milky Way really are. Fortunately, there are many conservation societies throughout the world that share this agenda, not forgetting the very important Dark Sky societies specifically aimed at reducing light pollution, and we as astronomers must promote their aims and agendas.

The night sky and the Milky Way are truly wonders of the Universe we reside in and are part of the place we call home, the Earth. They have been our companions since humans first looked up towards the stars many thousands and perhaps millions of years ago, and yet in just a few generations we may lose them. We need to, nay must, keep these wonders, not just for us, but for all people for all time. So please become aware that we are losing, slowly but surely, our access to the night sky. Join the conservation societies that actively promote safe and efficient night time lighting, and become an active member of the Dark Sky Association. Show your family and friends how amazing the night sky is and convince them that it must not be obscured any further.

As was once sung in a song, "we are star stuff", and we as astronomers, and in fact all people, would do well to remember this.

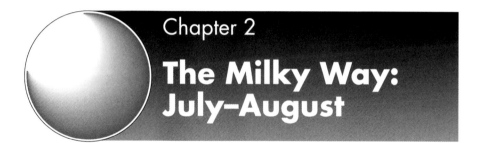

Chapter 2

The Milky Way: July–August

Sagittarius, Serpens Cauda, Scutum, Aquila, Hercules, Sagitta, Delphinus, Vulpecula, Cygnus, Lyra, Lacerta

RA 18^h to 23^h; Dec. –40° to + 60°; Galactic longitude[1] 0° to 110° ; Star Chart 2

2.1 Sagittarius

Let me state now that there are those amateur astronomers who believe this part of the Milky Way is *the* most spectacular, and who am I to disagree[2] (see Star Chart 2.1)? The star clouds of Sagittarius are justifiably some of the most wonderful and awe-inspiring regions that can be observed (see Figure 2.1). But before you grab hold of your binoculars and make a mad dash outside, there is of course another side to this. For those lucky observers who live in, say, southern Europe or the southern United States, and those who are very fortunate to live in equatorial regions, then these skies will provide views and scenes you are unlikely to ever forget. Those of us, however, who live in northern Europe and Canada have to deal with the unfortunate fact of life that Sagittarius will always lie close to the horizon, and sometimes when we read about the amazing sights that await observers in this region of the sky and then try to see them we are often left with a sense of disappointment. The only advice I can offer is this: these regions are truly spectacular, so try to observe with an unobstructed horizon, and with dark skies. If this is not possible, book a holiday to a location where the skies are clear and Sagittarius is at the zenith, You will never forget it![3]

This constellation, which incidentally transits in early July, is packed full of emission and dark nebulae, open and globular clusters and superb star fields. For those observers who may only have binoculars, take heart, because even with just simple equipment one can spend many evenings just scanning the regions of this part of the Milky Way (see Star Chart 2.1). In fact, it should come as no surprise to you that we could devote a whole section of the book just to Sagittarius.

---

[1] See Appendix 1 for details on astronomical coordinate systems.
[2] Well, actually, I think the Milky Way in Cygnus along with its mysterious Great Rift is equally wonderful.
[3] I was once fortunate enough to observe from the depths of Turkey where the sky was unbelievable. I was able to see many objects in the Milky Way, with the naked eye, that I had only ever read about. The moral of this tale is simple – dark skies are crucial to observing faint objects!

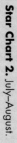

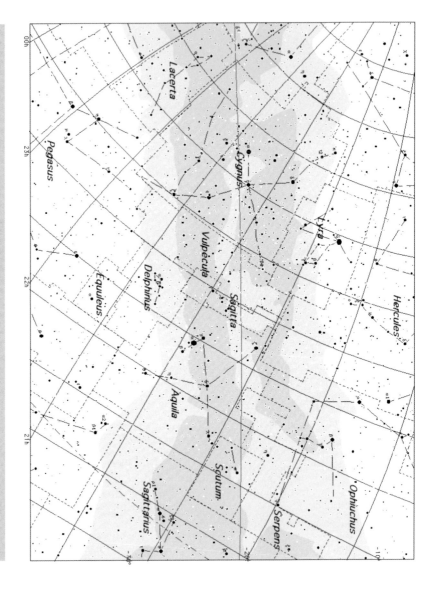

Star Chart 2. July–August.

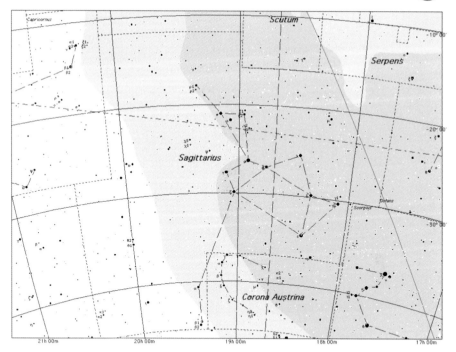

Star Chart 2.1. Sagittarius.

Ignoring for a moment the plethora of splendid objects, the most important fact about Sagittarius, as it relates to this book, is that the center of the Milky Way lies within it.[4] The center is actually about 4° northwest of **Gamma (γ) Sagittarii** (see Star Chart 2.2). The center of the Galaxy has always posed several problems to astronomers. Unfortunately, owing to the vast amount of gas and dust that lies between us, it has been impossible to see the center using visible light.[5] However, infrared and radio waves can escape from the center, and so a picture can be built up of the inaccessible region. What has been learnt is impressive. There is a radio source located very near or even at the exact center, called **Sagittarius A***,[6] and this was the first cosmic radio source discovered. Measurements of the radio source indicate that it is no bigger than the diameter of Mars's orbit. One of the surprising results is that Sagittarius A* is stationary, which would indicate that it is very massive.[7] The current estimate for the mass is about a thousand solar masses. All this information suggests that at the center of our Galaxy lies a black hole! This, of course, would mean that the actual center of our Galaxy must be invisible. Although, if you look through a telescope at the region, the field of view will be full of numerous star fields, that actually lie much closer to us than to the center. And when you are observing this region

---

[4] Bear in mind that the center isn't actually located in space in Sagittarius. It is just that when we look at the constellation we are looking towards the area where the center is located.

[5] It is estimated that the light is dimmed by about 27 to 28 magnitudes. This is a diminution of around 60 billion. That's a lot!

[6] Sagittarius A* is now believed to be made of two components; SgrA East and SgrA West. The former is a supernova remnant, and the latter is an ultra compact, nonthermal source, i.e., a black hole.

[7] Recent analysis suggests that the density around the center of the Galaxy is about a million times greater than any known star cluster. It is probably made up of stars, dead stars, gas and dust, and of course a black hole.

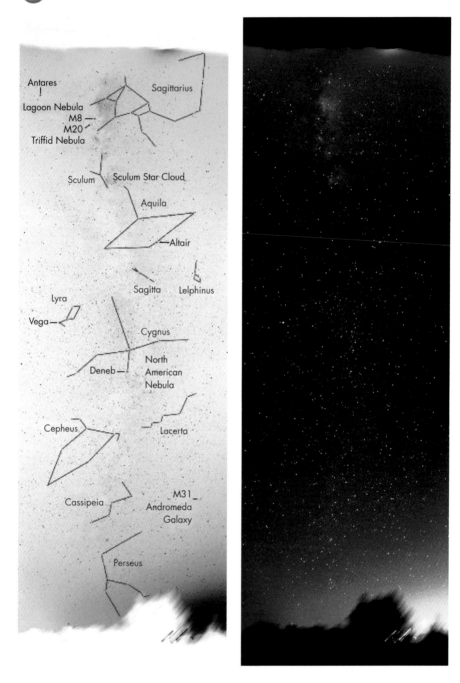

Figure 2.1. The summer Milky Way (Thor Olson).

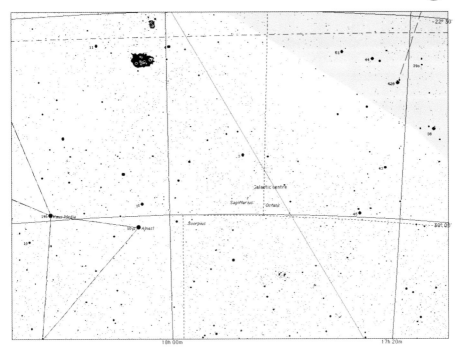

Star Chart 2.2. Galactic center.

you may like to contemplate the almost certain fact that even though you cannot see the center, there in your eyepiece, ever invisible, lies a supermassive black hole, around which you and I, the Solar System and the Galaxy rotate.

The Solar System is located about 30,000 light years from the center on the inner edge of the Orion–Cygnus Spiral Arm. What this means is that when we observe Sagittarius we are looking across an interarm gap that is more or less empty,[8] towards the next spiral arm that is closer to the galactic center. This arm is the **Sagittarius–Carina Spiral Arm** and within it are the majority of Messier objects. Exceptions to this are Messier 23 and Messier 25, both open clusters that are actually in the interarm gap between the two spiral arms.

We are lucky enough to be able to see through a gap in the dark dust clouds of the Sagittarius–Carina Spiral Arm and peer even deeper into our Galaxy's interior and observe a star cloud of the **Norma Spiral Arm.** This very rich star cloud is called the **Small Sagittarius Star Cloud.** But that is about as far as we can see into the inner regions of out Galaxy. I always wonder, when I scan this part of the sky, what hidden gems lie forever beyond our sight. Perhaps one day, far, far in the future, we may find out.

We can also observe something called the **Great Sagittarius Star Cloud.** This is a region of the sky that extends in a northerly direction from the stars **Gamma (γ) Sagittarii** and **Delta (δ) Sagittarii.** This is part of the central hub of the Galaxy that happens to bulge out from beneath the dark dust (see Figure 2.2).

Let's now concentrate on some of the individual objects located in Sagittarius, and as is usual, we will start with several stars.

---

[8] There are of course a few old stars and open clusters.

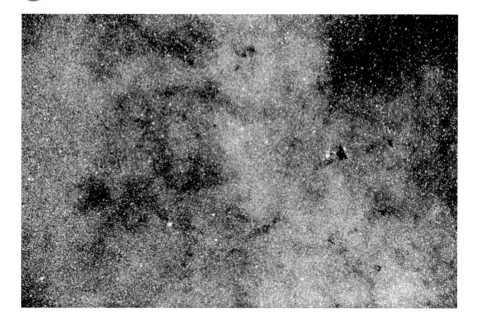

Figure 2.2. The Great Sagittarius Star Cloud (Mario Cogo, www.intersoft.it/galaxlux).

A nice double star is **h (Herschel) 5003**. This consists of a pair of reddish-orange stars of magnitude 5.2 and 6.9, although there are reports of some observers seeing an orange-yellow pair. They are separated by about 5.5 arcseconds and lie in a wonderfully rich star field. There is incidentally a very faint 13th magnitude star some 23 arcseconds distant that can be glimpsed under very good conditions, but will need a 25 cm telescope. A rigorous test for small telescopes is **21 Sagittarii**, and even a challenge for medium apertures. This is a nice contrast of orange and blue stars, although some observers report the secondary as greenish. They are at magnitudes 4.9 and 7.4 respectively. Another double is **HN 119**, first mentioned in William Herschel's catalogue of 1821. This is a nice double of orange and blue stars at magnitudes 5.6 and 8.6.

A rather rare type of star is **AQ Sagittarii,** which is classed as a carbon star. It is a bright irregular variable that can be seen in a 15 cm telescope. What makes it a nice star is that it is a deep red, as most carbon stars usually are, and shows a nice contrast with a faint white star to its northwest.

A fine triple star for small telescopes is **54 Sagittarii**. This is a wonderfully colored star system, with a yellow-orangeish primary, a pale blue secondary and a pale yellow companion. They have magnitudes of 5.4, 11.9 and 8.9 respectively. Let's finish by looking at a quadruple star, **Eta (η) Sagittarii**. It consists of a pair of stars of unequal magnitude, showing an orange primary with a close white secondary, magnitudes 3.2 and 7.8 respectively. There are two other stars making up the system: a 10th magnitude star some 90 arcseconds distant and a 13th magnitude star some 33 arcseconds distant. The system should be easily seen in a telescope of, say, 15 cm aperture and a medium magnification.

Now let's begin to look at those objects for which this part of the Milky Way is rightly famous – the nebulae and clusters!

Our first object is **Messier 23 (NGC 6494)**. Shining with a magnitude of 5.5, this cluster is often overlooked because it lies in an area studded with celestial showpieces. It is a wonderful cluster that is equally impressive seen in telescopes or binoculars, but the latter will only

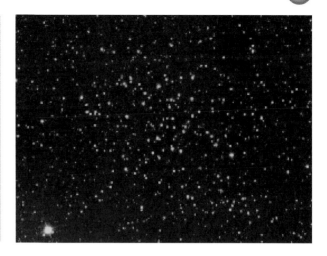

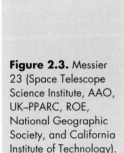

Figure 2.3. Messier 23 (Space Telescope Science Institute, AAO, UK–PPARC, ROE, National Geographic Society, and California Institute of Technology).

show a few of the brighter stars shining against a misty glow of fainter stars (see Figure 2.3). It has about 100 members and covers an area of around 30 arcminutes. Like so many of its kind, it is full of double stars and star chains. A 10 cm telescope will show it nicely. A small cluster, only 7 arcseconds across, is **Herschel 7 (NGC 6520)** shining at magnitude 7.5. This cluster, although fairly bright, is situated within the **Great Sagittarius Star Cloud**, and thus makes positive identification difficult. It contains about three dozen faint stars and locating it is a test of an observer's skill. A further object that may interest you is the dark nebula Barnard 86, which is projected against the cluster and about which we will talk later.

An outstanding cluster for small telescopes and binoculars is **Messier 21 (NGC 6531)**. This is a compact, symmetrical cluster of bright stars with a nice double system of 9th and 10th magnitude located at its center (see Figure 2.4). It is also located very close to the Trifid Nebula. It is about 20 arcminutes across but because it is in a very rich region of the Milky Way I find it difficult to see where the cluster ends and the Milky Way begins! In the cluster is the grouping called **Webb's Cross**, which consists of several stars of 6th and 7th magnitude arranged in a cross. Several amateurs report that some stars within the cluster show definite tints of blue, red and yellow. Can you see them?

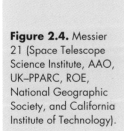

Figure 2.4. Messier 21 (Space Telescope Science Institute, AAO, UK–PPARC, ROE, National Geographic Society, and California Institute of Technology).

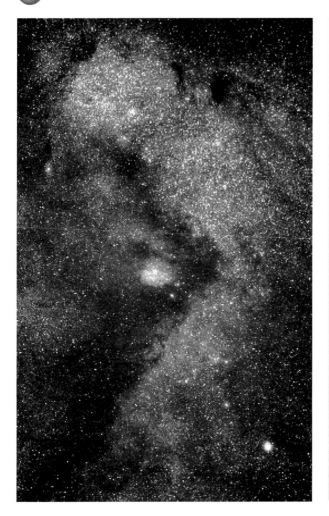

Figure 2.5. Messier 24. (SBAS)

Another superb object for binoculars is **Messier 24**, also known as the **Small Sagittarius Star Cloud**,[9] visible to the naked eye on clear nights at magnitude 2.5 and nearly four times the angular size of the Moon some 95 × 35 arcminutes. As mentioned earlier, the cluster is in fact part of the Norma Spiral Arm of our Galaxy, located about 15,000 light years from us. The faint background glow from innumerable unresolved stars is a backdrop to a breathtaking display of 6–10th magnitude stars (see Figure 2.5). It also includes several dark nebulae, which add to the three-dimensional impression. Many regard the cluster as truly a showpiece of the sky. Spend a long time observing this jewel! Located within the Small Sagittarius Star Cloud is the small cluster **NGC 6603**, which looks just like a globular cluster. It is small and nearly circular in shape, and will need a medium magnification in order to be resolved. Also located nearby is the dark cloud **Barnard 92** (see later).

At the other end of the scale is **Messier 18** (**NGC 6613**), a small and unremarkable Messier object, and perhaps the most often ignored. This little cluster about 10 arcminutes across, containing many 9th magnitude stars, is still worth observing. It is best seen with binoculars or low-power telescopes (see Star Chart 2.3). A double star is located within the cluster.

---

[9] It is also referred to as the **Little Star Cloud** in some books.

Visible to the naked eye, **Messier 25** (**IC 4725**) is a pleasing cluster suitable for binocular observation, as it shines at magnitude 2.6 and is about 32 arcminutes across. It contains several star chains and is also noteworthy for small areas of dark nebulosity that seem to blanket out areas within the cluster, but you will need perfect conditions to appreciate these. There are three nice deep-yellow stars arranged linearly in the center of the cluster (see Star Chart 2.4). The cluster is unique for two reasons: it is the only Messier object referenced in the **Index Catalogue** (IC), and is one of the few clusters to contain a Cepheid-type variable star – **U Sagittarii**. The star displays a magnitude change from 6.3 to 7.1 over a period of 6 days and 18 hours.

A nice cluster is **NGC 6645,** about 15 arcminutes in diameter and with around 70 stars ranging from 10th to 14th magnitude. It can be resolved as a cluster in a small telescope of, say, 8 cm, but is really best appreciated with a larger aperture.

This part of the Milky Way has a large number of globular clusters, many of them large and bright. Let's look at these next.

Our first is **NGC 6440,** which is relatively close to the Solar System, yet it experiences some four magnitudes of extinction due to dust (see Figure 2.6). It lies in a sparse part of the Milky Way, so that should tell you immediately that there must be a lot of dust about which is blotting out the star (see Star Chart 2.5). The globular is small, only 1 arcminute across, and so although you will be able to locate it with a telescope of 10 cm aperture, a much larger aperture will be needed to study it in any detail.

Located about 40 arcminutes northwest of Gamma Sagittarii is the small globular cluster **Herschel 49** (**NGC 6522**) (see Star Chart 2.6). It has a magnitude of around 8.6 and is 5.6 arcminutes across (see Figure 2.7). A 15 cm telescope will show a somewhat granulated aspect, but with telescopes of aperture 20 cm or more, this cluster will appear with a bright

Star Chart 2.3. Messier 18.

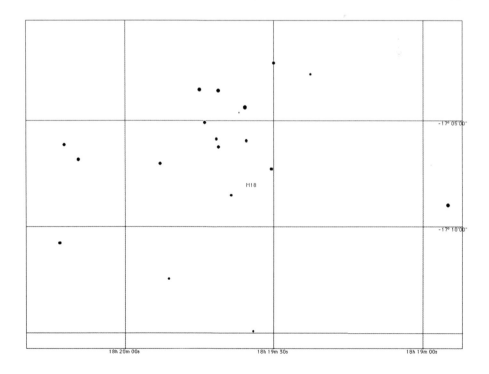

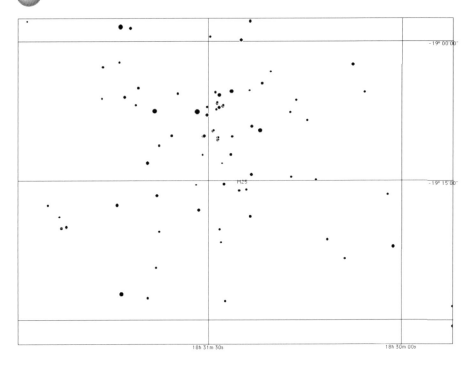

Star Chart 2.4. Messier 25.

core, but with an unresolved halo. It is, however, a difficult object to locate with binoculars. In the same field of view as NGC 6522 is **Herschel 200 (NGC 6528)**, a 9.5 magnitude globular. Even in a large telescope of aperture 35 cm, this cluster is unresolved (see Star Chart 2.6). It will just appear as a faint glow with a slightly brighter center. This is a particularly good challenge for large-aperture telescopes. What makes these two clusters nice is that even in a small telescope of, say, 8 cm, the difference between them is immediate. Try it and see.

Set deep within a most beautiful star field is the cluster **NGC 6544**. This is an irregularly round-shaped object about 1.5 arcminutes across (see Star Chart 2.7). With a small

Figure 2.6. NGC 6440 (Harald Strauss, AAS Gahberg).

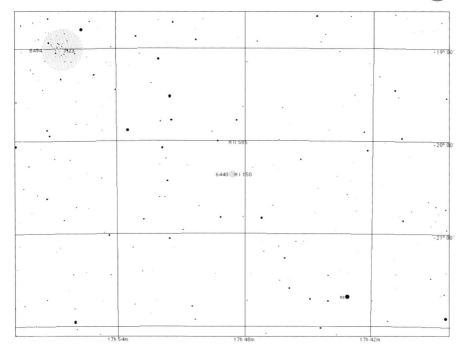

Star Chart 2.5 (*above*). NGC 6440.
Star Chart 2.6 (*below*). NGC 6522; NGC 6528.

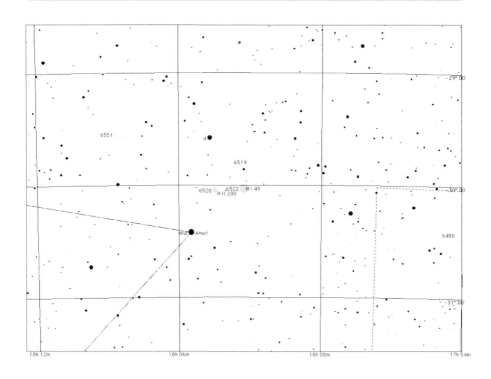

Figure 2.7 (*left*). NGC 6522 (Space Telescope Science Institute, AAO, UK–PPARC, ROE, National Geographic Society, and California Institute of Technology).

Star Chart 2.7 (*below*). NGC 6544; NGC 6553.

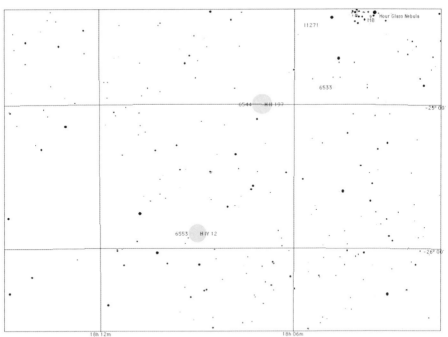

telescope of, say, 12 cm, a few stars around its periphery can be resolved, but a larger aperture will show many more, providing a high enough magnification is used.

A globular that always seems to look much better photographically than it does visually is **Herschel 12** (**NGC 6553**) an 8th magnitude cluster about 8.1 arcminutes in diameter (see Star Chart 2.7). Not easily visible in binoculars (although it would prove an observational

Figure 2.8. NGC 6553 (Space Telescope Science Institute, AAO, UK–PPARC, ROE, National Geographic Society, and California Institute of Technology).

challenge to locate), it is a fairly evenly bright cluster, with no perceptible increase towards the core (see Figure 2.8). A 10 cm telescope will find it easily enough.

An easy cluster to find in a telescope of even 8 cm aperture is **Messier 28** (**NGC 6626**). Shining at magnitude 6 and about 11.2 arcminutes across, this globular lies in a very nice, well-scattered star field. Only seen as a small patch of faint light in binoculars, this is an impressive cluster in telescopes (see Figure 2.9). With an aperture of 15 cm it shows a bright core with a few resolvable stars at the halo's rim. With a larger aperture the cluster becomes increasingly resolvable and presents a spectacular sight. It lies at a distance of about 22,000 light years. This is well worth seeking out for large-aperture telescope owners, as it is a lost gem.

When **Messier 69** (**NGC 6637**) was first discovered it was compared to the nucleus of a comet, and is a beautifully well-resolved cluster some 2.5 arcminutes across (see Star Chart 2.8). Visible as just a 7th magnitude hazy spot in binoculars, it appears with a nearly star-like core in telescopes (see Figure 2.10). A large aperture will be needed to resolve any detail, and will show the myriad dark patches located within the cluster.

Now for something special: the globular cluster **Messier 22** (**NGC 6656**). This is a wonderful, truly spectacular globular cluster, visible under perfect conditions to the naked eye

Figure 2.9. Messier 28. (SBAS)

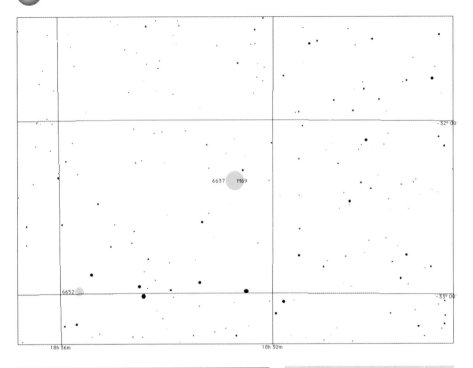

Star Chart 2.8 (*above*).
Messier 69.

Figure 2.10 (*left*). Messier 69
(Space Telescope Science
Institute, AAO, UK–PPARC, ROE,
National Geographic Society,
and California Institute of
Technology).

(see Star Chart 2.9). It is a bright 5th magnitude cluster spread over an area 24 arcminutes across, nearly the size of the full moon. Low-power eyepieces will show a hazy spot of light, while high power will resolve a few stars (see Figure 2.11). A 15 cm telescope will give an amazing view of minute bright stars evenly spaced over a huge area. It is often passed over by northern-hemisphere observers owing to its low declination, and this is a shame as it is a must-see object. It lies only 10,000 light years away, which is nearly twice as close as M13.

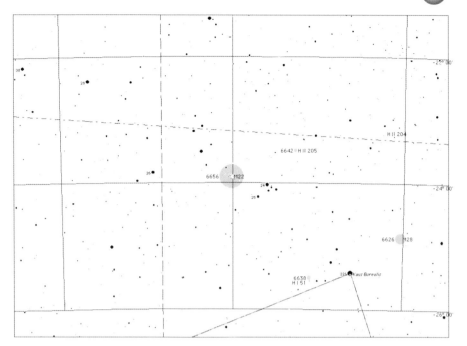

Star Chart 2.9 (*above*). Messier 22.
Figure 2.11 (*below*). Messier 22 (Harald Strauss, AAS Gahberg).

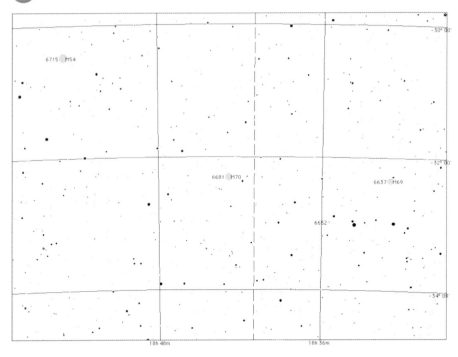

Star Chart 2.10. Messier 70.

A moderately bright, but not well-known cluster is **Palomar 8**, discovered by the great American astronomer G. O. Abell on the first Palomar Sky Survey. It can be seen in an 15 cm telescope but will appear plain and faint. Of course larger apertures will show more detail, and in a 30 cm telescope the resolution of stars will begin.

A faint globular that will need at least a 15 cm telescope for any resolution is **Messier 70** (**NGC 6681**) (see Star Chart 2.10). It is a faint 8th magnitude binocular object that is a twin of M69. It is best viewed with a large aperture, because with a small telescope, it is often mistaken for a galaxy (see Figure 2.12). It lies at a distance of 35,000 light years.

Discovered in 1778 is the cluster **Messier 54** (**NGC 6715**). With telescopic apertures smaller than 35 cm the cluster remains unresolved, and will show a view similar to that seen in binoculars, only larger – a faint hazy patch of light shining at a magnitude of 7.6. It has a colorful aspect – a pale blue outer region and pale yellow inner core (see Figure 2.13). Recent research has found that the cluster was originally related to the **Sagittarius Dwarf Galaxy,** but that the gravitational attraction of our Galaxy has pulled the globular from its parent. Among the globular clusters in the Messier catalogue it is one of the densest as well as being the most distant. It lies on the far side of our Galaxy, some 70,000 light years away – amazing!

The globular cluster **Palomar 9** (**NGC 6717**) would appear much easier to the observer if it weren't for the glare from the bright yellow components of **Nu (ν) Sagittarii** that lies some 2.5 arcminutes to its north (see Star Chart 2.11). It is a small cluster, only 30 arcseconds wide with a magnitude of about 9. For any detail to be seen, a large aperture will be needed, although it can be glimpsed in a 15 cm telescope.

A very fine globular is **NGC 6723**: a broadly compressed cluster with an irregular round shape, it can be seen in a 10 cm telescope with some of its stars becoming resolved (see Star Chart 2.12). It shines at about 7th magnitude.

Figure 2.12. Messier 70 (Space Telescope Science Institute, AAO, UK–PPARC, ROE, National Geographic Society, and California Institute of Technology).

Another object that was initially compared to a comet is **Messier 55** (**NGC 6809**), our final globular cluster (see Star Chart 2.13). This is a lovely cluster, easily seen in binoculars, and just visible with the naked eye at magnitude 6.3. Even in a telescope of 8 cm, it will show as a faint cloud of stars. Small-aperture telescopes (15 cm) will show a bright, easily resolved cluster with a nice concentrated halo (see Figure 2.14). Because it is very open, a lot of detail can be seen, such as star arcs and dark lanes, even with quite small telescopes. With a larger aperture, hundreds of stars are seen.

The Milky Way in Sagittarius is resplendent with both emission and dark nebulae, and has many of the most spectacular and colorful nebulae[10] that one can find anywhere. We look at some of these.

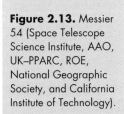

Figure 2.13. Messier 54 (Space Telescope Science Institute, AAO, UK–PPARC, ROE, National Geographic Society, and California Institute of Technology).

---

[10] The colorful nebulae that populate the pages of popular astronomy magazines and books do not often resemble anything that can be seen at an eyepiece. The reason is that the eye doesn't handle color too well at low light levels. Do not be disappointed, then, when after having seen the multicolored textures of, say, the Trifid Nebula, the real thing only shows a pale grey or perhaps bluish tint.

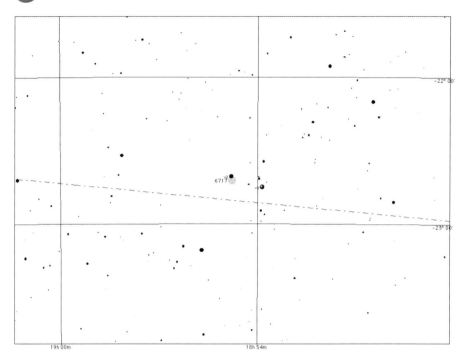

Star Chart 2.11 (*above*). NGC 6717.
Star Chart 2.12 (*below*). NGC 6723.

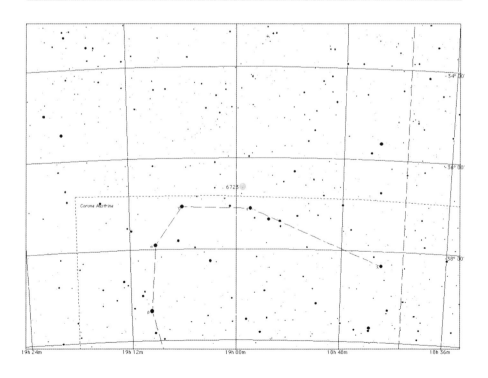

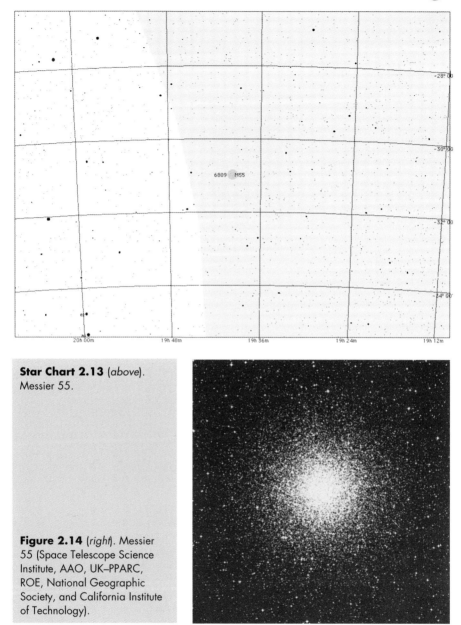

Star Chart 2.13 (*above*). Messier 55.

Figure 2.14 (*right*). Messier 55 (Space Telescope Science Institute, AAO, UK–PPARC, ROE, National Geographic Society, and California Institute of Technology).

One object that will need a very large field of view is **NGC 6476**. Actually, to be truthful, NGC 6476 is not one object, but rather a large patch of sky that contains innumerable star fields, and both bright and dark nebulosity. It is a region that benefits from sweeping with binoculars, and is a nice starter for what is to follow.

Our first port of call is the lovely nebula **Messier 20** (**NGC 6514**). It is also known as the **Trifid Nebula**. This emission nebula can be glimpsed as a small hazy patch of nebulosity, and in fact is difficult to locate on warm summer evenings unless the skies are very

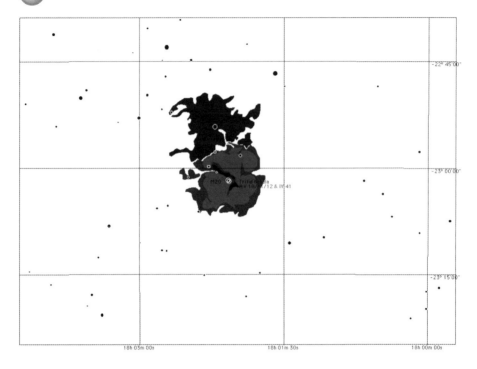

Star Chart 2.14. Trifid Nebula.

transparent (see Star Chart 2.14). With aperture around 15 cm, the nebula is easy to see, along with its famous three dark lanes that give it its name (see Figure 2.15). They radiate outwards from the central object, an O8-type star that is the power source for the nebula. The northern nebulosity is in fact a reflection nebula, and thus harder to observe. In large telescopes the nebula is wonderful and repays long and careful observation.

Often mentioned in the same breath as the Trifid nebula, although having no physical relationship, is the magnificent nebula **Messier 8** (**NGC 6523**), also known as the **Lagoon Nebula**. Visible to the naked eye on summer evenings, this is thought by many to be the premier emission nebula of the summer sky.[11] It is actually quite large, about 43 × 35 arcminutes, and will need a large field of view to be completely seen in one go (see Star Chart 2.15). Binoculars will show a vast expanse of glowing green-blue gas split by a very prominent dark lane (see Figure 2.16). Using a light filters and telescopes of aperture 30 cm will show much intricate and delicate detail, including many dark bands. The Lagoon Nebula is located in the **Sagittarius–Carina Spiral Arm** of our Galaxy, at a distance of 5400 light years. It also contains the bright open cluster **NGC 6530**. Preceding this cluster are two stars about 3 arcminutes apart. The brighter of the two is **9 Sagittarii**, and 3 arcminutes west and south of this star is the small but bright **Hourglass Nebula**.

A wonderful emission nebula is **Messier 17** (**NGC 6618**), also known as the **Swan Nebula** or **Omega Nebula** (see Star Chart 2.16). This is a magnificent object in binoculars, and is

---

[11] I was never able to see this object with the naked eye from the UK, but it is an easy object from most parts of the USA, and it really is spectacular.

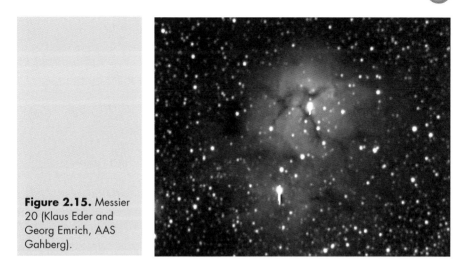

Figure 2.15. Messier 20 (Klaus Eder and Georg Emrich, AAS Gahberg).

perhaps a rival to the Orion Nebula, M42, for the summer sky (see Figure 2.17). It is very bright and is roughly 15 × 3 arcminutes. Alas, it is not often observed by amateurs, which is a pity as it offers much. This is what the well-known British astronomer Paul Money has to say about the nebula: "M17 has to be a favorite as it was one of the earliest nebula that I

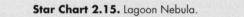

Star Chart 2.15. Lagoon Nebula.

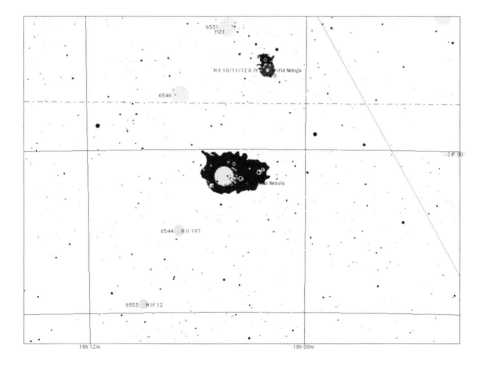

Figure 2.16 (*above*). Messier 8 (Harald Strauss, AAS Gahberg).
Star Chart 2.16 (*below*). Swan Nebula.

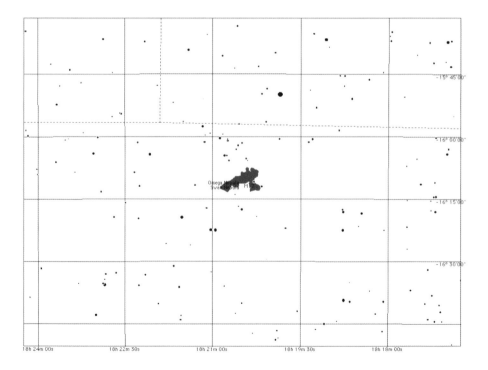

Figure 2.17. Messier 17 (Klaus Eder and Georg Emrich, AAS Gahberg).

found using a humble 60 mm refractor in my early years of observing. Even in that small instrument it had a distinctive shape – a tick against the night sky – and it makes me smile when I revisit it with any instrument. I always feel that God was pleased with himself and with his work and placed the tick amongst the stars for us to see." I couldn't have said it better myself. With telescopes the detail of the nebula becomes apparent, and with the addition of a light filter it can in some instances surpass M42. Certainly, it has many more dark and light patches than its winter cousin, although it definitely needs an [OIII] filter for the regions to be fully appreciated. This is another celestial object that warrants slow and careful study.

One of the most difficult types of nebulae to observe are reflection nebulae, and fortunately for us Sagittarius has its fair number of them. Two examples are **NGC 6589** and **NGC 6590**. These are located on the southern edge of the Small Sagittarius Star Cloud. The southernmost nebula surrounds the pair of stars **h (Herschel) 2827** that are both at 10th magnitude. There are also several patches of dust that obscure the stars, but occasionally a clear area is found, and 15 arcminutes northeast of these reflection nebula is such an area, where a bright star excites its surrounding gas to give rise to the faint emission nebula **IC 1283–84**. The star in question is actually **Beta (β) 246**, and can be seen in a 30 cm telescope.

It is strange to recall that a class of nebulae that is very common in Sagittarius and the Milky Way as a whole, and yet does not glow, are the dark nebulae. Rather they are conspicuous by their absence of light. Let's look at some of these dark nebulae. Our first is **Barnard 289**, which is outlined against the profuse star fields. With an irregular shape it's around 30 ×15 arcminutes in size, lying roughly north–south, and can be seen to gradually merge with the surrounding clouds of stars. A few stars can be seen within it of course, but the usual luminescent background of the Milky Way is absent. A similar situation occurs with **Barnard 87**. It is also known as the **Parrot Nebula**. Although not a very distinct nebula, it stands out because of its location within a stunning background of stars. A few bright stars lie to its south, and in its center is a lone star of magnitude 9.5. Visible in binoculars as a small circular dark patch, it is best seen in small telescopes of around 10–15 cm aperture.

An object situated in the Small Sagittarius Star Cloud is **Barnard 92**. It lies on the star cloud's northwestern edge and is some 15 arcminutes across. On dark and transparent nights it can be seen as an obvious hole in the star-filled backdrop. Another dark nebula, this time in the Great Sagittarius Star Cloud, is **Barnard 86**, also known as the **Ink Spot**.

This is a near-perfect example of a dark nebula, appearing as a completely opaque blot against the background stars.

Our penultimate group of objects are planetary nebulae. There are quite a few of these in Sagittarius, but most are small and very faint, and so we will discuss just a representative few.

A pale gray planetary nebula some 22 arcseconds northeast of the globular cluster NGC 6440, **NGC 6445** is a small and faint object, less bright and half the size, of the afore-mentioned cluster (see Star Chart 2.17). It can be seen, barely, in a 10 cm telescope, but a larger aperture is really needed for this (see Figure 2.18).

Another planetary that this time shows a definite bluish tint is **NGC 6565**. It is small, and will appear about 10 arcseconds across in a telescope of 30 cm or more. In a smaller telescope, say 15 cm aperture, it will only be about 5 arcseconds across (see Star Chart 2.18). Anything smaller and it will appear stellar-like. An [OIII] filter makes an appreciable difference. Located some 20 arcminutes to its southwest is the dark nebula **Barnard 90**.

Located within a dense part of the Milky Way is **NGC 6563** (see Star Chart 2.19). It appears well defined and disk-like, with a pale blue tint, some 45 arcseconds across. It can be glimpsed, if the conditions are right, with a telescope as small as 8 cm. A difficult nebula to locate, but once found you will think worth the effort, is **NGC 6567**. It is within the Small Sagittarius Star Cloud, and is only 8 arcseconds across, but is fairly bright and somewhat hazy. Its appeal lies with the innumerable stars that surround and embrace it (see Star Chart 2.20).

Another planetary that was discovered by William Herschel, although he did not recognize it for what it was, is **NGC 6629**. It is a pale grey in color, some 15 arcseconds across. In

Star Chart 2.17. NGC 6445.

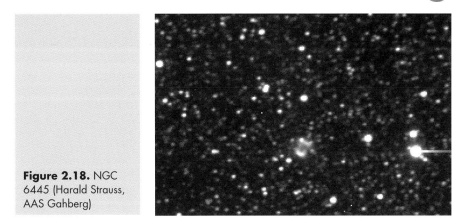

Figure 2.18. NGC 6445 (Harald Strauss, AAS Gahberg)

a telescope of, say, 20 cm, it will appear stellar if a low magnification is employed. Using a higher power, its true nature is apparent. Under superb conditions, its central star, magnitude 12.8, can be seen.

Our final planetary nebula, and penultimate object in Sagittarius, is **NGC 6818**, often called the **Little Gem**. It is a nice grey-blue object and can be seen in a telescope as small as 8 cm, although a 20 cm telescope will show its disk (see Figure 2.19).

Star Chart 2.18. NGC 6565.

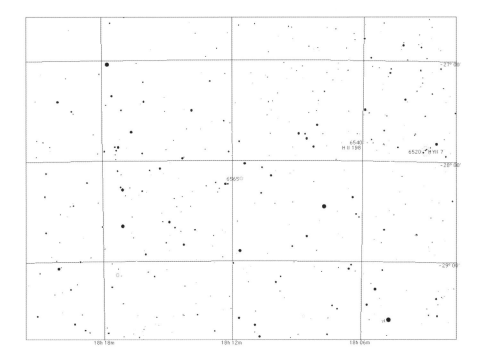

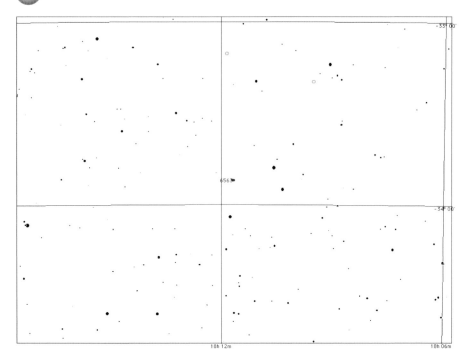

Star Chart 2.19 (*above*). NGC 6563.
Star Chart 2.20 (*below*). NGC 6567.

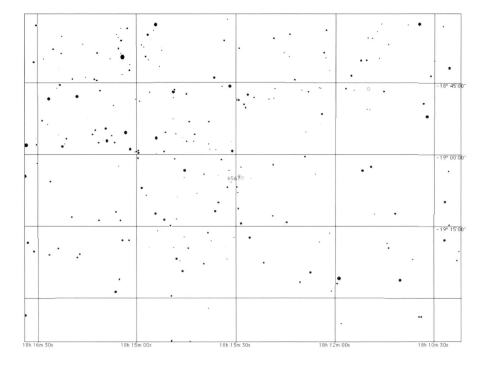

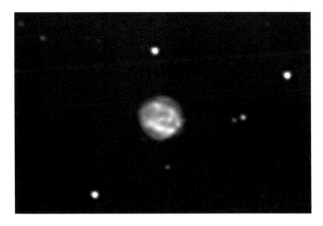

Figure 2.19. NGC 6818 (Robert Schulz, AAS Gahberg).

Our final object with which we leave the wonderfully rich region of the Sagittarius Milky Way may be a surprise to you. It is a galaxy, and a famous galaxy at that. It is **NGC 6822**, also known as **Barnard's Galaxy (Caldwell 57)** (see Figure 2.20). This is a challenge for binocular astronomers because even though it is fairly bright at magnitude 8.8, it has a low surface brightness and so is difficult to locate, but it can be glimpsed, even in small binoculars, barely, as a indistinct north–south smear (see Star Chart 2.21).

Once found, however, it will just appear as a hazy indistinct glow, running east–west, some 20 × 15 arcminutes. This is in fact the bar of the galaxy. Strangely enough, it is one of those objects that seems to be easier to find using a small aperture, say 10 cm, rather than a large one. Nevertheless, dark skies are essential to locate this galaxy, as its poor and feeble light tends to merge with the multitude of foreground stars. It is an irregular galaxy and a member of the **Local Group** of Galaxies. If you do manage to find it, you should congratulate yourself, as it is something of an achievement to do so.

Figure 2.20. NGC 6822 (Space Telescope Science Institute, AAO, UK–PPARC, ROE, National Geographic Society, and California Institute of Technology).

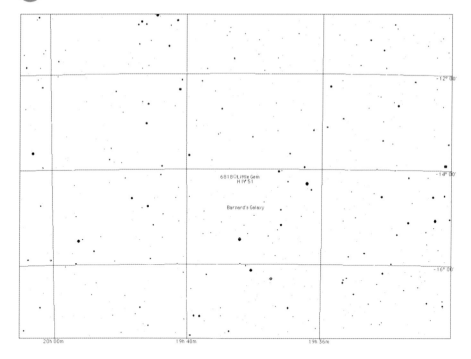

Star Chart 2.21. NGC 6822; Barnard's Galaxy.

2.2 Serpens Cauda

We are now going to look at a small constellation, Serpens Cauda (see Star Chart 2.22), which in fact is the lower half of a constellation that is split into two by Ophiuchus. To be precise, the constellation actually transits towards the latter part of June, and so if we were going to be strict, it should be in the accompanying volume. However, it lies next to the constellation Sagittarius (see above) and no doubt you will be scanning both constellations at the same time, so I deemed it wiser to put it here.

To the naked eye it also looks as if the Milky Way is only in the southern part of Serpens Cauda, and if you look at older star atlases the same situation will apply, but in fact the complete constellation is immersed within the Milky Way and lies very close to the galactic equator, so that is the approach we will take here, as it will allow us to observe far more objects. Even though Serpens Cauda lies within the Milky Way it is subject to quite a bit of obscuration by dust, and the keen-eyed observer will notice immediately that most of the little constellation is covered by the **Great Rift** – a very dark and immense dark dust cloud that actually divides the Milky Way into two. It begins in Cygnus and ends in Centaurus.[12] But this doesn't mean that there are no nice objects within in Serpens Cauda. Far from it – it has its fair share of clusters and stars, and one magnificent nebula. So let's start by looking at a few stars.

---

[12] Towards the northern aspect of the constellation there is a particularly dark dust cloud near the open cluster IC 4756 (see later) that blots out the light from the Milky Way.

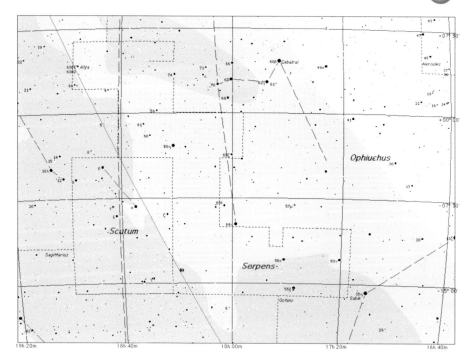

Star Chart 2.22. Serpens Cauda.

A nice double to begin with is **5 Serpentis**, but on closer observation you will see that in fact it is a triple system of unequally bright stars, which always seems to enhance color contrast. It consists of a pair of yellow and red stars with a faint white companion, with magnitudes 5.1, 10.1 and 9.1. It lies about 22 arcminutes southeast of Messier 5 (see below) and a 10 cm telescope should have no difficulty resolving the pair.

A lovely colored pair is **Nu (ν) Serpentis**. This has the quite rare colors of green coupled with blue. It is a very wide pair, separation 47 arcseconds, with magnitudes of 4.3 and 8.3. This is a star system that is well worth observing. A couple that has no color contrast, but nevertheless is nice to observe is **Delta (δ) Serpentis**. They are pale yellow in tint and are separated by nearly 3 arcseconds with magnitudes 5.26 and 4.22. It is a nice double for small telescopes.

A star that is situated in a part of the Milky Way that is threaded throughout with dark nebulae is **Σ (Struve) 2303**. It is a very close double and so something of a test for the small telescope, consisting of a yellow and white star, magnitudes 6.6 and 9.1 respectively, separated by 2.1 arcseconds. You will notice straight away the dearth of field stars. A test for larger telescopes is **AC (Alvin Clark) 11**. This pale yellow pair of nearly equal magnitude stars are only 0.8 arcseconds apart, and in some circumstances with a high enough magnification will appear as two disks just touching each other. At least a 20 cm telescope will be needed for this object.

Another system that is set in a very dark part of the Great Rift is **59 Serpentis**. Colored white and yellow it is set in an almost blank part of the sky, which looks quite amazing when you consider the surrounding star fields. An 8 cm telescope will do well here.

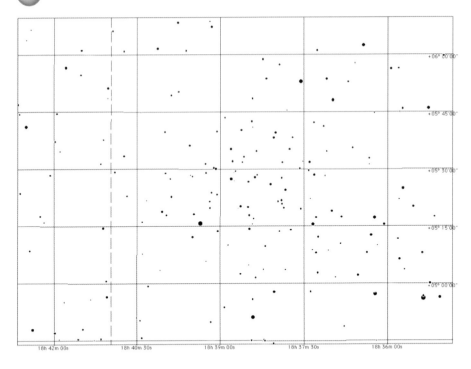

Star Chart 2.23. IC 4756.

Our final objects are a couple of very strongly colored blue stars, **Theta (θ) Serpentis**. With magnitudes of 4.69 and 5.06, and separated by 22.3 arcseconds, they are easily resolved, and can be split even in moderately sized binoculars.

There are several open clusters in Serpens Cauda, but the majority are faint and so we will look at just a few. With a diameter of nearly 1°, **IC 4756** is a splendid object for small telescopes and binoculars (see Star Chart 2.23). There are over 80 stars ranging from 7th to 9th magnitude, and such is the integrated magnitude of all these stars that the cluster can, under good conditions, be glimpsed with the naked eye. With the magnificent backdrop of the Milky Way, it is a lovely object to observe, but do not use too high a magnification as otherwise the clustering effect will be lost.

We now come to the most famous and important object in the constellation, the open cluster, **Messier 16 (NGC 6611)**, and its attendant nebula, **IC 4703**, also known as the **Eagle Nebula** or the **Star Queen Nebula** (see Star Chart 2.24). A fine large cluster easily seen with binoculars, Messier 16 has an integrated magnitude of about 6. It is about 7000 light years away, about 40 light years in diameter and is located in the Sagittarius–Carina Spiral Arm of the Galaxy (see Figure 2.21). Its hot O-type stars provide the energy for the Eagle Nebula, within which the cluster is embedded, and can be seen as a hazy glow surrounding the stars. We are looking at a very young cluster of only 800,000 years, with a few of its members at an even younger 50,000 years old.

The nebula IC 4703 is a famous though not often observed nebula. Although it can be glimpsed in binoculars, and will appear as a hazy patch with the naked eye, telescopic observation is needed to see any detail. As is usual, the use of a [OIII] filter enhances the visibility. The "**Black Pillar**" and associated nebulosity are difficult to see, even

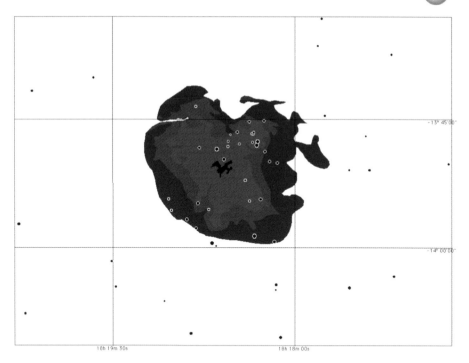

18h 19m 30s   18h 18m 00s

Star Chart 2.24 (*above*). IC 4703.
Figure 2.21 (*below*). Messier 16 (Klaus Eder and Georg Emrich, AAS Gahberg).

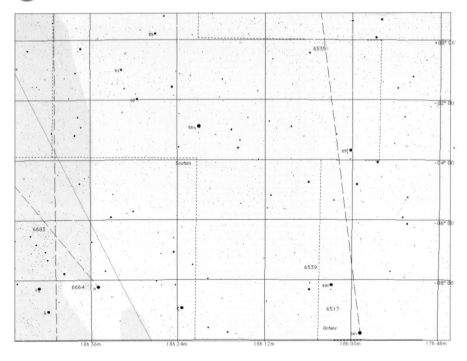

Star Chart 2.25. NGC 6535; NGC 6539.

though they are portrayed in many beautiful photographs.[13] Nevertheless, they can be spotted by an astute observer under near-perfect conditions, especially if contrast-enhancing filters are used. The nebula is about 66 by 54 light years in size and may be, along with Messier 17, the Swan Nebula in Sagittarius, part of a much larger nebula complex.

Finally, there are two globular clusters we can look at: **NGC 6535** and **NGC 6539** (see Star Chart 2.25). The former is a small cluster about 1 arcminute across (see Figure 2.22). A telescope of 15 cm aperture will only show a round, dim hazy spot with perhaps one or two stars. Interstellar absorption is high here, but research has shown the cluster is inherently faint. The latter cluster is, however, very heavily obscured by dust and so is a faint object. It can just be seen with a 15 cm telescope but will never be resolved, even with telescopes as large as 30 cm (see Figure 2.23). What will be apparent is the near absence of field stars, as compared to other regions, which is due to the dust. Try observing this area and then immediately going to one of the rich star clouds in Sagittarius, and you will see the effect the dust can have.

2.3 Scutum

We are now going to look at a small constellation. However, it is one in which the Milky Way of summer shines in all its glory and also contains some truly spectacular objects, but

---

[13] A prime example of astronomical imagery fooling the amateur into thinking that these justifiably impressive objects can easily be seen through a telescope.

Figure 2.22. NGC 6535 (Harald Strauss, AAS Gahberg).

more about those as we go on (see Star Chart 2.26). **Scutum,** for that is how it is known, is a relative newcomer to the list of constellations, having been created in 1690, and as its brightest stars are of only 3rd magnitude, it is a difficult constellation to locate visually. It transits in early July.

Let's begin by looking at a few variable stars. **R Scuti** is a semiregular variable star, of the **RV Tauri** type. It exhibits a varying period that averages about 140 days varying from 5th to 6th magnitude. On occasion it can brighten to magnitude 4.9, and at other times drop to 8th magnitude. Studies have also shown that it has a secondary cycle of about 1300 days. It is a nice deep-yellow star located about 1° due south of **Beta (β) Scuti**. Incidentally, RV Tauri stars are not confined to the Milky Way, but can be found in globular clusters and the Galaxy's central bulge. Another variable star, which is the prototype of its class, is **Delta (δ) Scuti**. These are short-period pulsating variable stars with a small magnitude change. In fact, in this particular case the change is too small to be detected visually. They are believed to be related to Cepheid variable stars, perhaps a low-mass relation.[14] A bonus for us is that the star is also a triple star system. The primary is a nice pale yellow and the companion is a strong blue of magnitudes 4.7 and 9.2 respectively. A superb triple star is Σ **(Struve) 2306**, with magnitudes of 7.9, 8.6 and 9.0 and separated by 10.2 arcseconds. This is

Figure 2.23. NGC 6539 (Harald Strauss, AAS Gahberg).

---

[14] Other examples of this class of variable star are **Rho (ρ) Puppis** and **Beta (β) Cassiopeiae**.

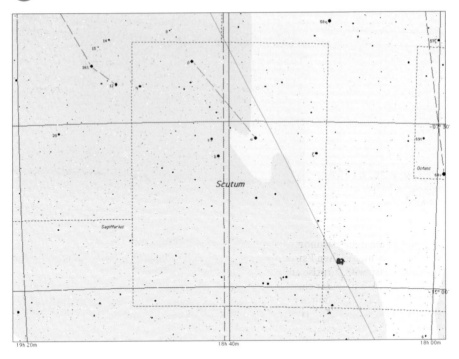

Star Chart 2.26. Scutum.

a wonderful triple star system of delicately colored stars. Observers have reported the primary as gold or copper-colored and the secondary as cobalt blue or blue. The blue secondary will need a high magnification in order to split it, but do try to observe this gem, as the colors are lovely. Our final star is the nice double **Σ (Struve) 2373**. It consists of a pair of yellow stars, one pale, and the other deep. Easily seen with a telescope of about 8 cm, it lies in a lovely star field.

Now let's look at some star clusters, including one that is a favorite of mine. There are, incidentally, over 10 open clusters in Scutum, but most are faint so we shall concentrate just on those that can be seen with moderate instruments. Our first is **Messier 26 (NGC 6694)**. Shining at an integrated magnitude of around 8, this is a small but rich cluster containing 11th and 12th magnitude stars, set against a haze of unresolved stars (see Figure 2.24). This makes it unsuitable for binoculars, as it will be only a small hazy patch of light, and so apertures of 10 cm and more will be needed to appreciate it in any detail. It lies less than a degree from Delta Scuti and spans around 15 minutes of arc (see Star Chart 2.28).

The next cluster, however, is a true celestial showpiece: **Messier 11 (NGC 6705)**. Also known as the **Wild Duck Cluster**, this is a truly delightful object. Although it is visible with binoculars as a small, tightly compact group, reminiscent of a globular cluster, they do not do it justice (see Figure 2.25). With telescopes, however, its full majesty becomes apparent. Containing many hundreds of stars, it is a very impressive cluster. It takes high magnification well, where many more of its 700 members become visible. At the top of the cluster is a glorious pale yellow tinted star. The British amateur astronomer Michael Hurrell called this "one of the most impressive and beautiful celestial objects in the entire sky", and I tend to agree with him. I always show the Wild Duck Cluster to people who

Figure 2.24. Messier 26 (Harald Strauss, AAS Gahberg).

may not be amateur astronomers, first with a low magnification. This doesn't bring forth much comment, but then I switch to a high magnification. This usually does the trick as they often gasp with surprise and awe! (See Star Chart 2.28.)

There is also a globular cluster we can look at – **NGC 6712**. This is a moderately bright cluster about 2.5 arcminutes across, and with a 20 cm telescope some resolution may be achieved (see Figure 2.26). It can be seen with a telescope as small as 10 cm, but only as a hazy spot. What is special though is that the planetary nebula **IC 1295** is only about 0.5° away and so can be seen in the same field of view (see Star Chart 2.27). A somewhat faint object about 1.5 arcminutes across, it can be glimpsed in a 20 cm telescope, but is a

Figure 2.25. Messier 11 (Robert Schulz, AAS Gahberg).

Figure 2.26. NGC 6712 (Harald Strauss, AAS Gahberg).

challenge in anything smaller. Of course a [OIII] filter makes everything easier. The central star becomes visible only in the largest telescopes. Both the globular cluster and the planetary nebula are situated in a glorious star field.

It will not come as much of a surprise to know that there are no galaxies to be seen here, because of the thick obscuring dust, and that may also be the reason that emission and reflection nebulae are so scarce. However, there is something visible to us, and that is the reflection nebula **IC 1287**. This is a large, but faint object, illuminated by the star **Σ (Struve) 2325**. It will need a large telescope in order to be observed.

Star Chart 2.27. NGC 6712; IC 1295.

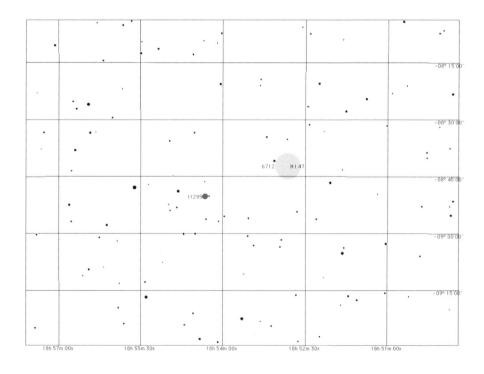

Figure 2.27. The Scutum Star Cloud. (SBAS)

Now for a truly wonderful object. I have mentioned that the Milky Way completely covers the constellation, but there is one part of the Milky Way itself that warrants a mention, and this is the **Scutum Star Cloud**. This is a mélange of very dark dust clouds and very bright star clouds (see Figure 2.27). The dark clouds are themselves part of the larger Great Rift, mentioned earlier, that crosses the northern part of the constellation. The Scutum Star Cloud lies between **Alpha (α) Scuti** and **Beta (β) Scuti** and is one of the most wonderful sections of the entire Milky Way. When we look at this object we are in fact looking towards the interior of our Galaxy. Try using binoculars or a rich-field telescope and you will be rewarded with a field filled with barely resolved stars without number.

Here is what the American astronomer Jim Mullaney has to say about the star cloud: "The sky's largest deep-sky wonder. Wondrous clusters … many astral splashes in this crowded district of the Galaxy. And here – as with the great big billowy star clouds of Scorpio, Sagittarius & Cygnus – watch for an amazing "3-D" effect that can occur without warning: as the eye–brain combination makes the association that the fainter stars you're seeing in the Cloud are farther away than are the brighter ones – that you're actually looking at layer upon layer of stars – the Milky Way can suddenly jump right out of the sky at you in a striking illusion of depth-perception!"

The edge of the Great Rift can be easily seen making inroads to the star fields. In some areas there will be lone outposts of stars, while in other regions it will be the dark clouds that are surrounded by stars. Such is the plethora of dark clouds that several have been given individual Barnard numbers. Of these **Barnard 103** and **Barnard 110** are particularly striking. Barnard 103 is easily seen at the northeast edge of the Scutum Star Cloud. It is a curved dark line and covers nearly 40 arcminutes. Try seeing a star along it! It can be glimpsed in binoculars, but is best seen at apertures of around 10–15 cm. Barnard 110 is also an easily seen complex of dark nebulae that can be seen in binoculars. The contrast between the background star clouds and the darkness of the nebulae is immediately seen. There are also many other dark nebulae; for instance, **Barnard 111** and **Barnard 119a** are close to and surrounding the open cluster Messier 11. The former, which is north of M11, is in fact a part of the Great Rift that extends south down into the Scutum Star Cloud. Incidentally, both Barnard 103 and Barnard 113 are the darker parts of the larger cloud Barnard 111. Then there is **Barnard 318** that lies south of M11, and **Barnard 112** lying even further south, which is nearly 20 arcminutes across. With all these dark nebulae, a dark sky that is also very transparent will be necessary. This is a wonderful constellation for observing with binoculars and letting not only your eyes but your imagination roam though the star clouds.

2.4 Aquila

We now come to a constellation that tends to group observers into two camps. There are those who think **Aquila** is a lovely summer constellation that is perfect for scanning with binoculars and wandering about its many Milky Way star fields and star lanes and chains, and there are those who think it is a disappointing constellation with few objects to grab an observer's attention. I actually reside in the former camp, but you must make up your own mind! The constellation is fairly high in the sky, about 45° above the horizon for mid-northern latitudes and transits in mid-July[15] (see Star Chart 2.28).

Star Chart 2.28. Aquila.

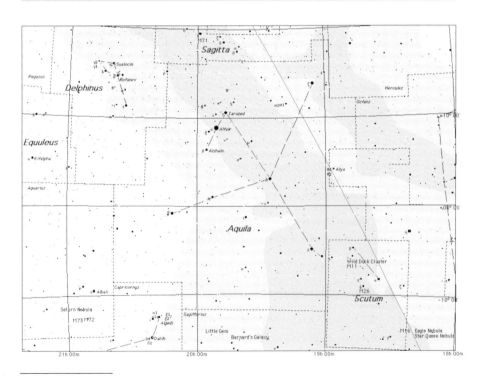

---

[15] The easternmost regions of the constellation are not actually within the Milky Way, and in older star atlases only the central region is located within it.

It has a few open and globular clusters, several quite good dark nebulae, and a lot of nice double and triple stars. There are even some galaxies present, but to be honest these are small and faint. The reason why there are so few clusters to be seen is that the Great Rift passes through the constellation, in which there are many large clouds of obscuring dust. They are about 500–1000 light years away and block out the light from the clusters that may lie in that particular direction. This does mean, however, that there are a lot of dark nebulae to observe. There is also one very strange object in Aquila, but I think I will leave that till last. However, it does have the lovely star **Altair**, which is a good star to begin with.

Alpha (α) Aquilae, the twelfth-brightest star at magnitude 0.76 is a lovely pure white colored star, although some observers see a hint of yellow. Flanked above and below by **Beta (β) Aquilae** and **Gamma (γ) Aquilae**, it is one of the famous **Summer Triangle** stars, the other two being **Deneb** and **Vega**. It also has the honor of being the fastest-spinning of the bright stars, completing one revolution in approximately 6.5 hours. Such a high speed deforms the star into what is called a flattened ellipsoid, and it is believed that because of this amazing property the star may have an equatorial diameter twice that of its polar diameter. It is believed to be at a distance of 16.77 light years away from us. A star that has a very deep red color is **V Aquilae**. This is a semiregular variable star, of average magnitude 7.5, with a period of about 350 days. It ranges in brightness from 6.48 to 8.1 magnitude and is a carbon star type C5. Another very fine red star is **R Aquilae**, which also varies irregularly in brightness over a period of some 284 days. Located within the Great Rift, it has a very large magnitude range, going from a brightish 5.5 to a faint 12.0 magnitude when it appears at its reddest. It is one of the Mira-type variable stars.

Another star that should be observed is the Cepheid variable **Eta (η) Aquilae**, located in the outer regions of the Milky Way. It ranges in magnitude from 4.1 to 5.3 every 7.2 days, and thus is easily within the range of small binoculars and even, under good conditions, the naked eye.

Two nice double stars that show color contrast are **11 Aquilae** and **23 Aquilae**. The former is a pair of 5.2 and 8.7 magnitude stars that have a nice yellow-blue contrast, whereas the latter[16] shines at magnitudes 5.31 and 8.76 and has a lovely color contrast of deep yellow and greenish blue. Another nice double is **Σ (Struve) 2404**, which is a nice pair of orange stars of 7th and 8th magnitude separated by about 3.6 arcseconds. They are located within a splendid star field and the brighter of the pair is itself a spectroscopic binary. One thing to note is that the colors have been reported as being yellow and blue, which may be due to the chromatic aberration in the telescopes used. Observe and see what colors you find.

Now for a test of both your eyes and telescope optics. The star **Zeta (ζ) Aquilae** is a brilliant white star that has a 12th magnitude companion, but the glare from the primary is so strong that you will need at least an aperture of 25 cm or larger to glimpse it, as well as very dark and transparent skies. Another test, although for a smaller telescope of, say, 8 cm aperture, is **Chi (χ) Aquilae**. It is a deep yellow or golden star of magnitudes 5.8 and 6.68. It has a very close 0.5 arcseconds separation and so is difficult to resolve.

Our next double star is **Pi (π) Aquilae**, which is another deep-yellow system, with magnitudes of 6.47 and 6.75, and with a separation of 1.4 arcseconds it is a little easier than the previous two entries to resolve. With some instruments it may appear as if the two stars are in contact. The penultimate double star is a very easy object to resolve and should be split with an 8 cm telescope. It is **Σ (Struve) 2587**. With a separation of nearly 4 arcseconds and at magnitudes 6.7 and 9.4 it is a nice object. Finally we have **β 57**, a lovely orange star set amongst a wonderful star field. The companion star is faint and located to its southeast and should be visible, providing the conditions are good, in an 8 cm telescope.

---

[16] There is in fact a very faint 13.7 magnitude star some 12 arcseconds away that will make the system a triple star system, but it is only visible in large telescopes.

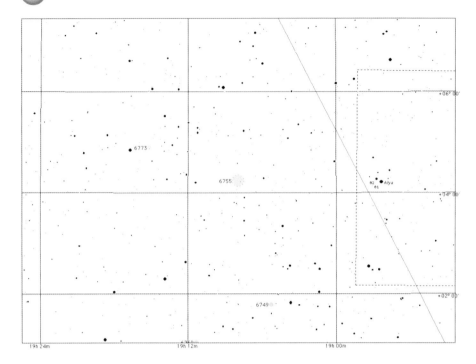

Star Chart 2.29. NGC 6755.

There are a few star clusters we can observe in Aquilae, but none are what we could call impressive. Nevertheless, they are worth seeking out. A small but nice open cluster of about 30 stars is **NGC 6709** (see Star Chart 2.30). Set in a field of about 15×12 arcminutes, the individual stars are, alas, too faint to be seen, and all that can be glimpsed is a pale haze. This of course means that it is a difficult object to locate. Nevertheless it is an object to persevere with. The cluster **NGC 6755** may be the easiest and nicest open cluster to see in Aquila (see Star Chart 2.29). It is an irregularly shaped object, and fairly conspicuous in small telescopes of, say, 10 cm aperture that stands out well from the Milky Way. With larger telescopes, more than 60 stars of 10th magnitude and fainter can be seen. Larger apertures will resolve even more members with nice star arcs and chains and many splendid doubles on view.

Another open cluster is **NGC 6738**. Like its predecessor above, it too is faint but a few 9th magnitude stars can be glimpsed in binoculars along with the ever-present haze of unresolved cluster members (see Star Chart 2.30).

We will end our look at open clusters with a couple that should appeal not only to owners of large telescopes but also, perhaps, for their uniqueness. They are **NGC 6773** and **NGC 6795**. Both of these clusters are faint and sparse and will need telescopes of at least 30 cm for any worthwhile details to be seen. However, what makes them interesting for us is that they are examples of what are known as nonexistent clusters in the revised NGC Catalogue.[17] The former has about 30 stars ranging from 12th to 14th magnitude, whereas the latter has some 60 stars with a few at 9th magnitude, but most are 11th and fainter.

---

[17] Other clusters in Aquila that are cataloged as nonexistent are NGC 6828, 6837, 6840, 6843 and 6858.

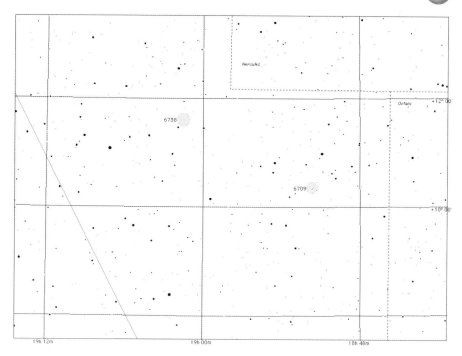

Star Chart 2.30. NGC 6709; NGC 6738.

A few globular clusters demand our attention (see Star Chart 2.31). The first of these is **NGC 6749**, a faint object in an area of heavy dust obscuration. It is about 2 arcseconds across and has a low surface brightness. Small telescopes will not be able to resolve much and large telescopes won't fare much better. Next is **NGC 6760.** Shining with a magnitude of 9.1, this is a faint, symmetrical cluster with a just perceptible brighter core (see Figure 2.28). It is about 2 arcminutes across and high-power binoculars should be able to locate this cluster. Even in a small telescope it should present no problems. But knowledge of the use of setting circles would be useful, as would a computer-controlled telescope. Our final globular is another test for those of you with large instruments. It is **Pal (Palomar) 11.** It is a faint and sparse object that will look like an indistinct pale haze about 1.5 arcminutes across with perhaps just two or three stars within it. It is just seen in a 25 cm telescope, providing the conditions are excellent.

As I mentioned earlier, there is a lot of dust and dark clouds in Aquila, and amongst the most famous and easily seen are **Barnard 142** and **Barnard 143.** Located about 3° north-west of Altair, this is an easily seen pair of dark nebulae, visible in binoculars. Covering an area some 80 × 50 arcminutes, it appears as a cloud with two "horns" extending towards the west. The nebula contrasts very easily with the background Milky Way and so is a fine object. With a rich field telescope and large binoculars, the dark nebula actually appears to be floating against the star field. If you have never tried observing one of these dark clouds, try this one, as I think you will be pleasantly surprised. Another dark cloud is **Barnard 133,** which is somewhat smaller than the two mentioned above at only about 10 arcminutes in diameter. It will look like a definite hole some 2° south of Lambda Aquila. As with all dark nebulae, dark clear nights and a complete absence of, say, light pollution will be a prerequisite for observation.

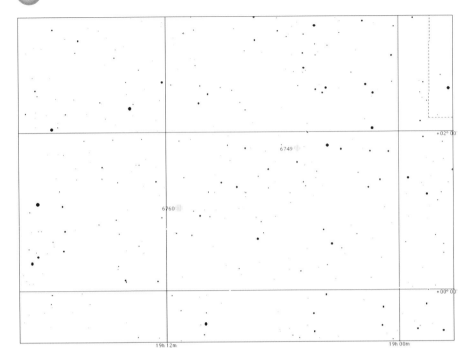

Star Chart 2.31. NGC 6749; NGC 6760.

Oddly enough there is one class of object that can be found in plenty in Aquila, and these are planetary nebulae. Many are faint and small, but there are a few nice examples we can look at. Our first is **Sh (Sharpless) 2–71 (PK 036–1.1)** (see Star Chart 2.32). Lying in a heavily obscured region, this faint planetary is about 1.5 arcminutes across, and surrounds a 13.5 magnitude central star (see Figure 2.29). A 20 cm telescope equipped with a [OIII] filter may be able to find it, but a 30 cm telescope should have no trouble.

Figure 2.28. NGC 6760 (Robert Schulz, AAS Gahberg).

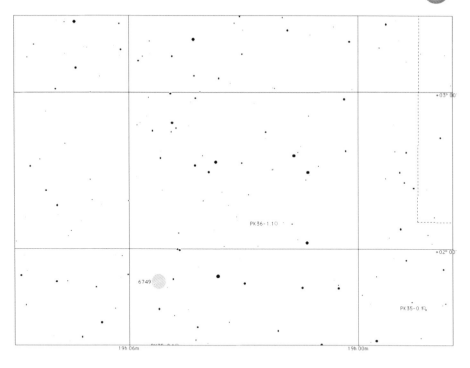

Star Chart 2.32. Sh 2–71.

Figure 2.29. Sh 2–71 (Space Telescope Science Institute, AAO, UK–PPARC, ROE, National Geographic Society, and California Institute of Technology).

Figure 2.30. NGC 6751 (Robert Schulz, AAS Gahberg).

Another nebula that can just be glimpsed in a telescope as small as 15 cm is **NGC 6751**. It appears greyish and is about 20 arcseconds across (see Figure 2.30). Larger apertures will be able to see its faint central star.

Another faint example is **NGC 6772** which lies in a nice star field. It has a low surface brightness and is around 1 arcminute in diameter, with no visible central star. Although it can be seen in a 20 cm telescope along with an [OIII] filter, a 30 cm aperture will naturally show more detail. One nebula that may prove difficult is **NGC 6778**. This is a small object, only 15 arcseconds across, and is quite faint. Even with a large telescope, not much is seen, and the central star can just be glimpsed, or imagined, in a 30 cm telescope. Incidentally, the dark nebula **LDN (Lynds) 619** is 10 arcminutes north of the nebula, and can be seen in outline against the field stars.

One planetary nebula that should present no problems is **NGC 6781 (Herschel 743)** (see Star Chart 2.33). This is an easily located planetary nebula – large, circular and bright (see Figure 2.31). Under excellent seeing, and using averted vision and dark adaption, a darkening of its center will be revealed along with the fainter part of its northern periphery. Large-aperture instruments will show far more detail, including the halo. The use of a [OIII] filter will help considerably and allow the granular morphology of the nebula to be glimpsed. Sadly, it is not visible in binoculars, although under excellent conditions it has been reported as visible in a 10 cm telescope.

Our final planetary nebulae are **NGC 6790** and **NGC 6803**. Both are very small and in all but the largest telescopes will appear star-like. The former, under the right conditions, has a pale blue disk and is best seen with the [OIII] filter, while the latter will only reveal itself under the highest magnification. Nevertheless, do try and seek these out.

Our final object in Aquila has for a long time given many headaches to astronomers. It is the strange and rather fascinating object known as **SS 433 (V1343 Aql)** (see Star Chart 2.34). This is a binary star that lies at the center of a supernova remnant, known as **W 50**. The binary cannot be resolved visually, but what makes this object so fascinating is that for a long time it remained an enigma as no one could really decide what it was. Its spectral lines are unique in some respects because they are shifted both to the red and the blue! This has subsequently been explained as due to jets of material emerging from a star with a velocity that is a staggering one-quarter that of the speed of light. As the star

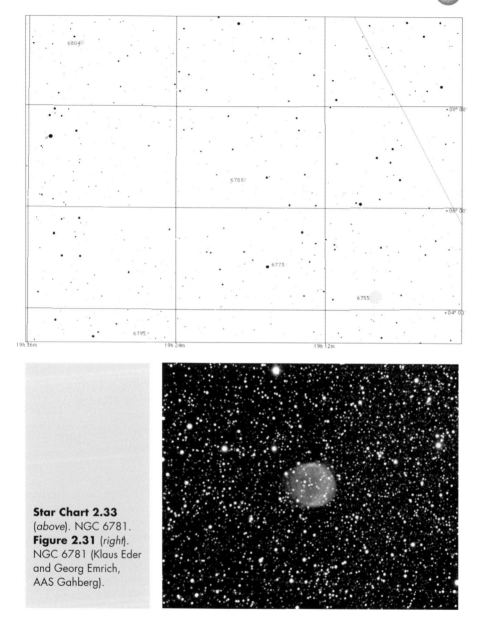

Star Chart 2.33
(*above*). NGC 6781.
Figure 2.31 (*right*).
NGC 6781 (Klaus Eder
and Georg Emrich,
AAS Gahberg).

rotates, or precesses, the jets sweep across the sky with a period of around 164 days. Further research indicated that whilst the primary of the binary system is an O (or maybe a B) type star, its invisible companion is a neutron star. The variations in brightness are very complex, but there are two main periods of 6.4 and 164 days. At maximum brightness, a telescope of 15 cm aperture will show the star as a faint point of light located amongst a rich star field. This is one of the Galaxy's most unusual objects and should be observed by everyone.

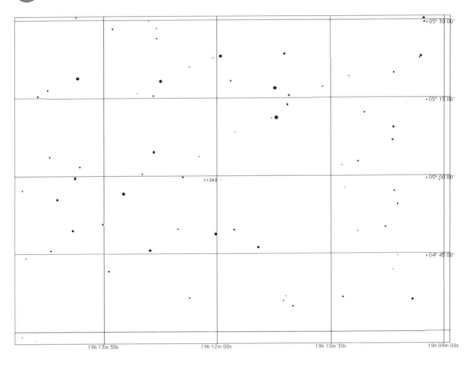

Star Chart 2.34. SS 433.

2.5 Hercules

Having this constellation in a book on the Milky Way may come as a surprise, but in truth its easternmost reaches do in fact have the Milky Way pass through them. This is something of a mixed blessing, as most of the lovely objects to be found in **Hercules** do not actually lie within the Milky Way areas (see Star Chart 2.35) Nevertheless, for the sake of completeness it must be mentioned. In addition, the center of the constellation actually transits in June, but the Milky Way region is more appropriately placed in this chapter as it transits in late July.

Having said that, there is only one object that is of interest to us as Milky Way observers, and that is **95 Herculis**. This is a famous pair, mainly due to the wide range of colors attributed to it. It has been described as anything from "apple green and cherry red", as Piazzi Smyth wrote in the 18th century, to both being pure white. Recent observations put the colors at a pale and a deep yellow. They are located in a nice star field with magnitudes of 4.93 and 5.31.

2.6 Sagitta

This little constellation is not referred to much, even though it has many splendid objects for us. I imagine that many observers have scanned the region of **Sagitta** in some detail and never realized it was there, having been awed and entranced by its bigger cousins to its north and south (see Star Chart 2.36). On warm summer nights this is a perfect constella-

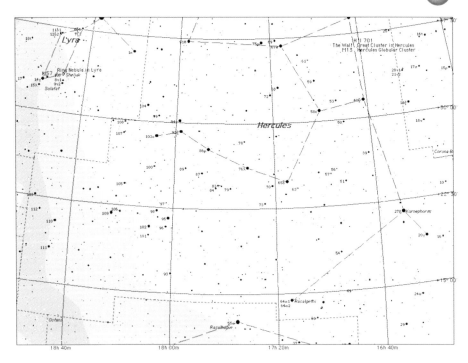

Star Chart 2.35. Hercules.

tion to scan with binoculars, as there are many star fields in which to lose oneself. It is located completely within the Milky Way and transits in mid-July.

Let's start our trip around this delightful constellation by looking at a few odd stars.[18] Our first is **WZ Sagittae** (see Star Chart 2.37). This belongs to a type of star known as a recurrent nova, which tends to flare up unexpectedly by as much as 10 magnitudes in a very short time, say several hours, then stays at its peak for a while before fading once again to its pre-outburst magnitude. There have been three major outbursts with WZ Sagittae, in 1913, 1946 and 1978. Normally the star is a very faint 15th magnitude, but can rise to 7th magnitude. In each of the aforementioned outbursts it only took a day to increase tenfold in magnitude, but about 60 days to fade again. Research indicates that it is a white dwarf star, and the increase in brightness results from an interaction with another unseen companion star. Some astronomers predict that the next outburst will be during 2010. So, all I can say here is "Watch the skies!"

Our next star is **FG Sagittae**, which is a very unusual variable star in that it is a pulsating variable of the **RV Tauri** class (see Star Chart 2.38). Over the past 80 years or so it has gradually increased in brightness from 13.7 magnitude to 9.5. What's more, this increase seems to have stopped, and now the star varies by about 0.5 magnitude over about two months. Deep imaging has revealed that the star is surrounded by a very tenuous nebula.

Another odd star is **V Sagittae**, which is an erratic variable star (see Star Chart 2.39). It varies in brightness between magnitude 9.5 and 13.9 and apparently has three overlapping periods of variability. This strange behavior seems to indicate that the star was recently, or is about to become, a nova, so this means it is worth watching in case an outburst is imminent.

---

[18] They are not really odd, just examples of stars that are relatively rare.

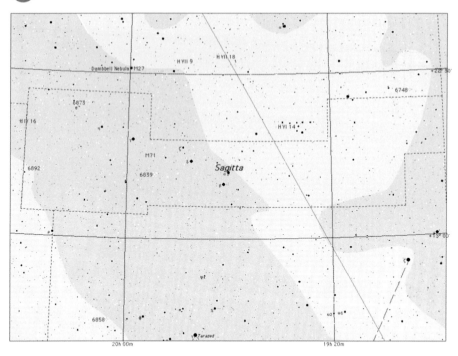

Star Chart 2.36 (*above*). Sagitta.
Star Chart 2.37 (*below*). WZ Sagittae.

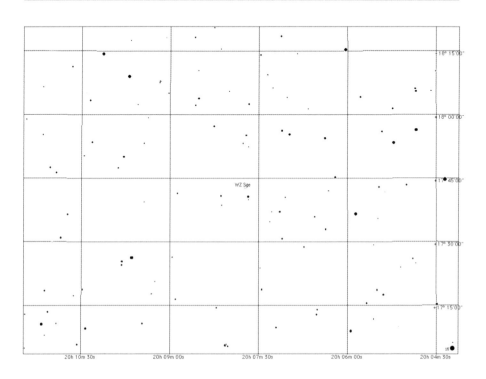

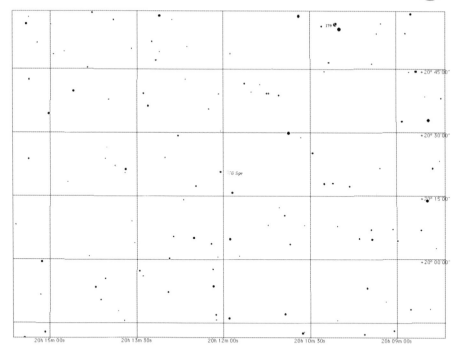

Star Chart 2.38 (*above*). FG Sagittae.
Star Chart 2.39 (*below*). V Sagitta.

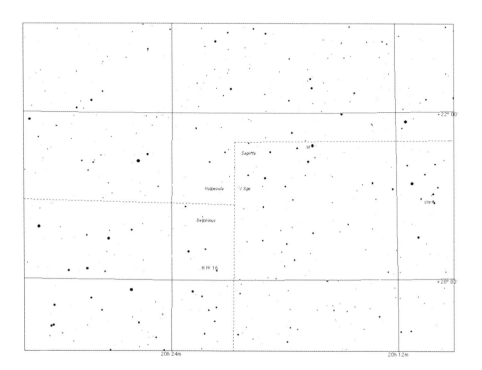

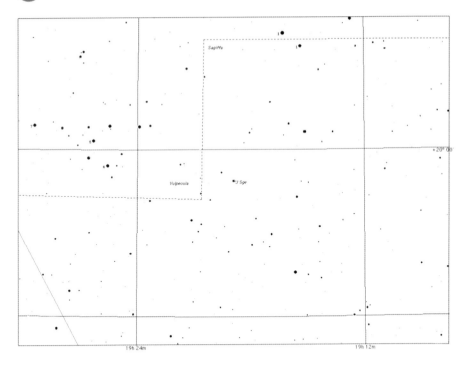

Star Chart 2.40. U Sagitta.

Our final star is **U Sagittae**, which is perfect for observing with binoculars (see Star Chart 2.40). It is an eclipsing binary star that varies with a period of 3.5 days, ranging in magnitude from 6.6 to 9.2 and back again. This variability is caused by the blue-white primary being eclipsed by the unseen yellow secondary.

There are some fine double and triple stars in Sagitta, and we shall now have a look at a few. A nice color contrast has been reported for **Zeta (ζ) Sagittae**, which can easily be seen in small telescopes. The primary is a bright 5.64 magnitude pale yellow color while the secondary is a 8.7 magnitude bluish star. Some observers believe the secondary is reddish! The bright primary is itself a very close binary that cannot be resolved by amateur telescopes. A fine triple star is **Theta (θ) Sagittae**. This is composed of two pale yellow stars with magnitudes 6.6 and 9.06, along with a orangeish companion, magnitude 7.4, some 85 arcseconds away. It is located on the southwestern edge of an open cluster, **NGC 6873**, which is one of the nonexistent clusters of the revised NGC Catalogue.[19]

There are also a couple of nice clusters we can look at, namely the open cluster Harvard 20 and the globular cluster Messier 71. A somewhat difficult binocular object, **Harvard 20** shines with an integrated magnitude of 7.7, but as the stars are of 12th and 13th magnitude, and spread out without any noticeable concentration, it is difficult to locate. The rich and compressed cluster, **Messier 71 (NGC 6838)**, will only appear as a very faint 8th magnitude glow in binoculars (see Figure 2.32). Located in a glittering star field, it is about 3 arcminutes across and through a small telescope it will not be resolved (see Star Chart 2.41). Up until recently there was some debate as to whether this was a globular or an open

---

[19] The coordinates for NGC 6879 as listed in the RNGC catalogue appear to be in error, as there is no cluster there. It is more likely that the cluster is the one that is close to Theta Sagittae.

Figure 2.32 (*above*). Messier 71 (Robert Schulz, AAS Gahberg).
Star Chart 2.41 (*below*). Messier 71.

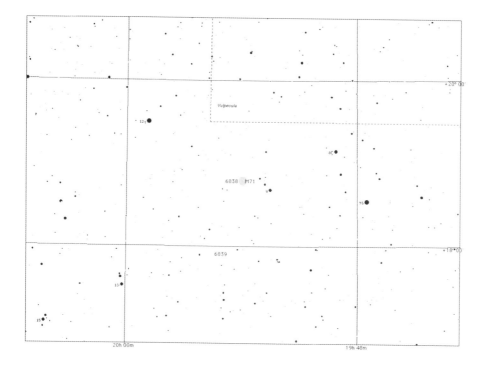

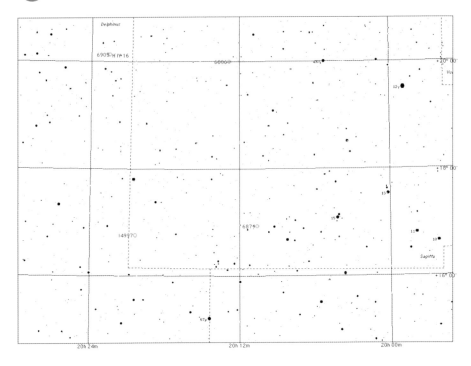

Star Chart 2.42. IC 4997; NGC 6879; NGC 6886.

cluster.[20] The consensus now is that it is a very young globular cluster only 9–10 billion years old, and only 13,000 light years away. What makes this globular so nice is that in a largish telescope, the central stars can be resolved all the way to the core, which is rare among globular clusters.

Finally, there are some planetary nebulae we can observe (see Star Chart 2.42). Set amongst a lovely star field is **IC 4997**, a small and bluish nebula which actually forms a rather nice double system with a yellow star that is about 1 arcminute southwest. It can be glimpsed under good conditions in an 8 cm telescope. It will, however, remain stellar in appearance even under the highest magnification. A very small planetary is **NGC 6879**, which can be used as a measure of your observing skills and, more likely, patience. It is about 5 arcseconds across and at magnitude 12.5, with an aperture of 15 cm, it will remain stellar. The orange double star **Σ (Struve) 2634** is about 14 arcminutes southwest, which may aid you. Or may not. Our final object is the planetary nebula **NGC 6886**. It is small, only 4 arcseconds across, with a magnitude of 11.4. It is slightly greenish and will need a moderate to high magnification to become resolved, otherwise it will just look like an out-of-focus star.

---

[20] It appeared that M 71 had more metals than was normal for a globular cluster and lacked the RR Lyrae type stars that are so typical for globulars. This has been explained by its relative youth – the stars have not evolved to the RR Lyrae stage of star evolution.

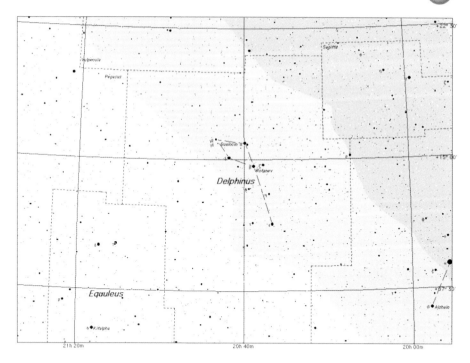

Star Chart 2.43. Delphinus.

2.7 Delphinus

We are now going to look at a delightful little constellation that sadly is often passed over by many amateurs: **Dephinus**. The Milky Way passes through its western regions and in fact completely surrounds the stars that give the constellation its name (see Star Chart 2.43). Sadly for us, neither of the bright globular clusters in Delphinus are in the Milky Way, so we shall not discuss them, but there are several other fine objects to observe. It transits at the end of July.

Fortunately, the two brightest objects in Delphinus are double stars, so let's begin by looking at them. The first is **Beta (β) Delphini**, which is a close binary with a period of 26.6 years and will be at its widest in 2004. When at their minimum separation the system looks pear-shaped in small telescopes. Both stars are white, although some observers state they can see a yellow tint to them, and are at magnitude 4.11 and 5.02. They are easily seen in an 8 cm telescope with a high magnification. The other nice double is **Gamma (γ) Delphini**, which is a beautiful system, consisting of a yellow primary and one of the rare green-colored stars as a secondary. The stars are separated by nearly 10 arcminutes and shine at 4.3 and 5.5 magnitudes respectively.

There is a nice variable star that can be observed in binoculars and small telescopes: **U Delphini**. It is a semiregular variable that varies in magnitude for about 6th to 8th over a period of 110 days. It lies about 2° north of Gamma Delphini.

Oddly enough, Dephinus has had quite a number of novae within its borders, and possibly one of the most famous in recent times was **HR Delphini**, also known as **Nova Delphini**

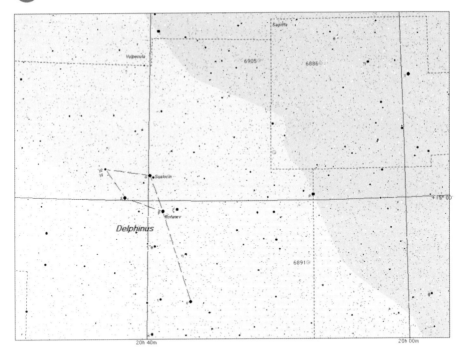

Star Chart 2.44. NGC 6891; NGC 6905.

1967. This lies about 3° north of the very distinctive kite-shaped asterism, and was discovered on 8 July 1967 by the late British astronomer George Alcock, when it was at magnitude 5.6. It then brightened to 4th magnitude in September, before fading slightly, and then brightened once again to magnitude 3.5 in December. By 1975 it had faded to magnitude 11.5. Now it is at 12th magnitude and resembles just one of the many Milky Way stars.

Small telescopes are ideal for observing two planetary nebulae in Delphinus although of course larger apertures will reveal more detail. The two nebulae are **NGC 6891 and NGC 6905** (see Star Chart 2.44). The former lies near the western edge of the constellation, some 1.5° southwest of a very distinctive pair of 6th magnitude stars. It has a magnitude of around 10.5 with a diameter of 14 arcseconds (see Figure 2.33). With a small telescope it will appear as a star-like object unless a high magnification is used, when it will then appear as a tiny disk. With larger apertures its true nature is revealed and a small, blue-green disk is seen. It has a 12th magnitude central star that can be glimpsed. The latter object, lying to the north of the constellation, is also known as the **Blue Flash Nebula**, and is a beautiful object. With a magnitude of 11.1 and with a diameter of nearly 40 arcseconds, it is lovely blue color (see Figure 2.34). In small telescopes a definite ring shape can be seen, with a sometimes-resolved central star. In larger apertures it is a truly lovely object and can be considered one of the better planetaries of the summer sky.

Even though the Milky Way is present here, open clusters and emission nebulae are conspicuously absent. There are, however, a couple of clusters we can look at, although one of them is more properly classed as an asterism. The open cluster **NGC 6950** is a very faint object, about 15 arcminutes across (see Star Chart 2.45). It is listed as one of the nonexistent open clusters we have discussed in earlier sections. At the position given for the cluster are about 35–40 12th magnitude stars.

Figure 2.33. NGC 6891 (Robert Schulz, AAS Gahberg).

The asterism **Harrington 9** is a little group of stars that also includes amongst them **Theta (θ) Delphini** (see Star Chart 2.46). It consists of a few 7th magnitude stars just to the east of Theta Delphini, surrounded by a further 25 stars of 8th and 9th magnitude stars. A very nice little object to conclude our visit to the constellation.

2.8 Vulpecula

Our next constellation is one that is little observed, mainly because it has no stars in it greater than magnitude 4.5. However, **Vulpecula** is absolutely swathed by the Milky Way (see Star Chart 2.47). In fact, it seems as if the constellation is split into two halves by it, as the Great Rift runs straight through it, and there is one very dark part of the Milky Way that lies near its western boundary.

Oddly enough, to the naked eye the Milky Way does not seem particularly bright here, but that viewpoint will change as soon as you look through, say, binoculars as there is

Figure 2.34. NGC 6905 (Klaus Eder and Georg Emrich, AAS Gahberg).

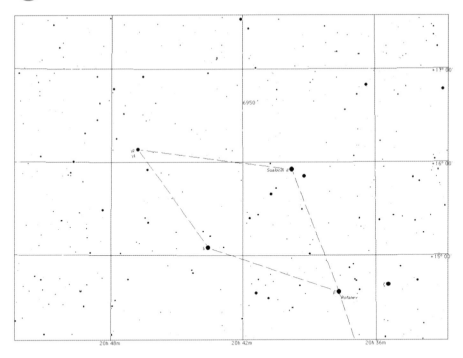

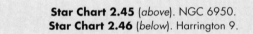

Star Chart 2.45 (*above*). NGC 6950.
Star Chart 2.46 (*below*). Harrington 9.

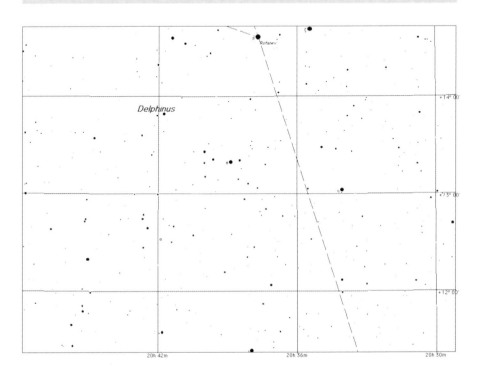

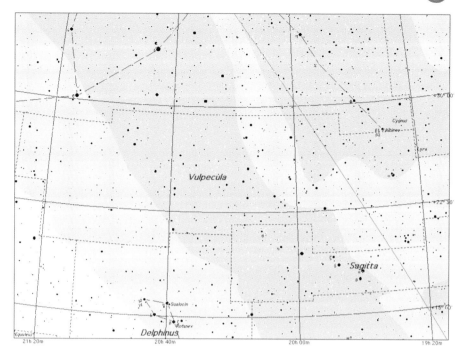

Star Chart 2.47. Vulpecula.

seemingly an infinite number of stars ranging in magnitude from 7th to 10th and fainter. As is to be expected, the blanketing effect of the Milky Way tends to blot out many of the galaxies, but there are several clusters and nebulae we can observe. It is an ideal constellation to sweep with binoculars on warm summer evenings. It is a wide constellation, encompassing over 2.5 hours of right ascension, and so it transits at the end of July.

There are several double and multiple stars in this part of the sky, yet few are what we could call spectacular. Most are close binaries that do not show much in the way of color contrast. However, there are two exceptions: β 441 and Σ (Struve) 2445. The former, also known as **Burnham 441**, is a nice pair, with a 6.2 magnitude primary and 10.7 secondary. Separated by nearly 6 arcseconds, it has a nice color contrast of yellow and blue. The latter is a lovely triple star that is perfect for binoculars and small telescopes, as it is separated by a wide 12 arcseconds, with magnitudes of 7.2, 8.9 and 8.9. It too presents a color contrast of blue and white stars.

There are some splendid open clusters here, so let's now look at some of these.

The first is **Collinder 399**, also known as either the **Coathanger Cluster** or **Brocchi's Cluster**. This delightful cluster[21] is often overlooked by observers, which is a shame as it is a large, dissipated cluster easily seen with binoculars; indeed, several of the brightest members, called **4, 5** and **7 Vulpeculae**, should be visible with the naked eye (see Figure 2.35). It spans over 1° of sky and contains a nice orange-tinted star and several blue-tinted

---

[21] Recent work suggests that it may not be a cluster at all, but rather an asterism.

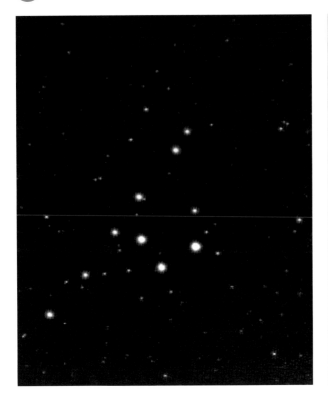

Figure 2.35.
Collinder 399. (SBAS)

stars. Its three-dozen members are set against a background filled with the Milky Way's fainter stars. Well worth observing during warm summer evenings.[22]

Another nice cluster is **Stock 1**. This is an enormous cluster that is best seen in binoculars, although it is difficult to estimate where the cluster ends and the background stars begin (see Star Chart 2.48). It appears that there are over 40 stars within a 1° area, although it may seem like it resembles a rich star field rather than a cluster. A somewhat difficult although interesting cluster is **NGC 6802**. This is elongated in shape, some 5 × 1.5 arcminutes, and in a small telescope will appear just as a longish smudge, although with larger apertures the stars become resolved (see Figure 2.36). It can be found at the eastern end of the Coathanger.

A cluster that is surrounded by a faint emission nebula is **NGC 6823** and **NGC 6820** respectively (see Star Chart 2.49). It lies about 6° southeast of Beta (β) Cygni (Albireo) (see Section 2.9 on Cygnus) and has over three-dozen members concentrated in a 6 arcminute area (see Figure 2.37). Many of the stars are tinted pale yellow, orange and blue. Under excellent seeing conditions you may be able to make out the nebula which extends to over have a degree around the cluster and an [OIII] filter may help.

Two clusters, **NGC 6882** and **NGC 6885** (**Caldwell 37**), are in reality just one cluster as it is difficult to distinguish one from the other. The former is about 18 arcminutes in size and the latter some 7 arcminutes. They are joined by a curved line of stars and seen in a wide field of view are a stunning object. Both can be observed in a telescope of 10 cm aperture, but of course a larger aperture, although decreasing the field, resolves more members (see Figure 2.38). As an aid to locating these faint objects, C37 is centered on the variable star **20 Vulpeculae.**

---

[22] Actually, it is worth observing at any time when conditions permit.

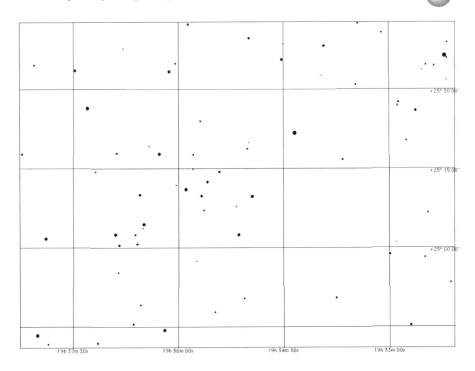

Star Chart 2.48 (*above*). Stock 1.
Figure 2.36 (*below*). NGC 6802 (Harald Strauss, AAS Gahberg).

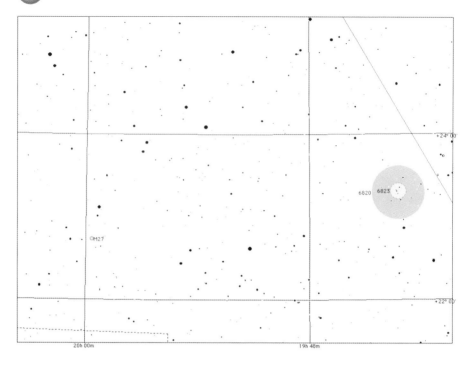

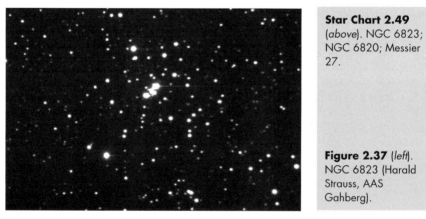

Star Chart 2.49
(*above*). NGC 6823;
NGC 6820; Messier
27.

Figure 2.37 (*left*).
NGC 6823 (Harald
Strauss, AAS
Gahberg).

Our final cluster is **NGC 6940**, which is best seen in a large field to really appreciate it. It is about 20×15 arcminutes in size, and seems to condense to an open central pattern where a bright orange star is located. With binoculars, only six or seven stars will be seen against the background haze of unresolved members (see Figure 2.39). One of the cluster's faint members is **FG Vulpeculae**, a red semiregular variable star that ranges in magnitude from 9.0 to 9.5 in about 80 days.

There are two planetary nebulae we can look at: one that is a test for the observing conditions and the telescope, and the other, possibly the best planetary in the sky. The former is **NGC 6842** (see Star Chart 2.50). It is only 50 arcseconds across and is faint at magnitude 13. It can be seen with a 30 cm telescope, but an [OIII] filter is really needed here. The latter object is the magnificent **Dumbbell Nebula**, or **Messier 27** (**NGC 6853**) (see Star Chart

Figure 2.38. NGC 6885 (Space Telescope Science Institute, AAO, UK–PPARC, ROE, National Geographic Society, and California Institute of Technology).

2.49). This famous planetary nebula, located south of 14 Vulpeculae, can be seen in small binoculars as a box-shaped hazy patch, and many amateurs rate this as the sky's premier planetary nebula (see Figure 2.40). In apertures of 20 cm, the classic dumbbell shape is apparent, with the brighter parts appearing as wedge shapes that spread out to the north and south of the planetary nebula's center. Here's what the British astronomer Peter Grego says of M 27, "I really enjoy viewing M27; it is the closest planetary nebula in the sky, and it can easily be picked up in 10 × 50 binoculars as a faint misty patch. In my 250 mm Newtonian there are more surprises – there's distinct mottling in the nebula and several faint stars can be discerned in the nebula's vicinity – these are foreground stars in the Milky Way that lie between us and the nebula". The central star can be glimpsed at this aperture, and it is also possible to discern some color. With perfect observing conditions, a faint glow can be seen in its outer parts. It is at magnitude 7.3 and is an enormous 6 × 4.5

Figure 2.39. NGC 6940 (Harald Strauss, AAS Gahberg).

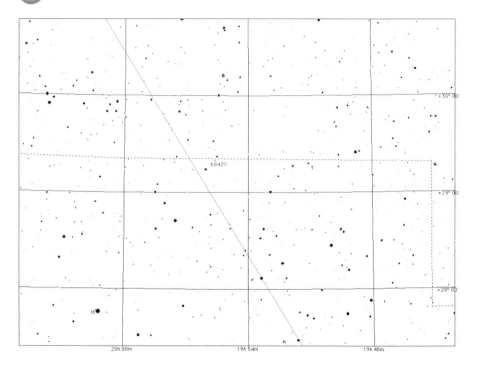

Star Chart 2.50 (*above*). NGC 6842.
Figure 2.40 (*below*). Messier 27 (Harald Strauss, AAS Gahberg).

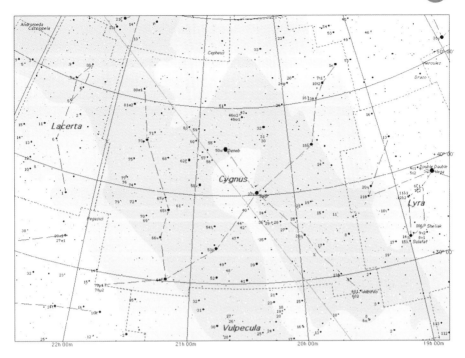

Star Chart 2.51. Cygnus.

arcminutes in size. This is truly a wonderful object and should be observed by all and every amateur astronomer.

2.9 Cygnus

Our next constellation of the autumn skies, **Cygnus**, is, in my opinion, one of the finest in the Milky Way. While it is true that none of its clusters and planetary nebulae are exemplary, what it does have is magnificent, and those are star clouds and vast fields of nebulosity set amongst truly spectacular views studded with gem-like stars (see Figure 2.41). Many observers agree that this is probably the finest part of the Milky Way for northern observers, riding high above us in the late summer skies (see Star Chart 2.51). It is low in the sky for southern observers and indeed some of it may be hidden from view. It transits at the end of July.

The splendid **Cygnus Star Cloud**, so obvious on summer and autumn evenings, is an incredible object in binoculars, stretching almost 20° from Albireo, Beta (β) Cygni to **Gamma (γ) Cygni**. It is in fact the brightest star cloud in the northern Milky Way. No matter what size or type of optical equipment is used to observe this feature, what becomes readily apparent is the bright glow from literally thousands of unresolved stars that fill the eyepiece. The next time you observe at this star cloud, just remember that you are looking

Figure 2.41. The Milky Way in Cygnus (Matt BenDaniel, http://starmatt.com).

down the length of the spiral arm we reside in just where it begins to curve in toward the interior of the Galaxy. Amazing!

The dark, and I think mysterious, **Great Rift** that splits the Milky Way here begins near Deneb and extends all the way down to the southern skies, ending near Alpha Centauri, and the dark nebulae called the Coalsack, discussed in detail in Book 2, is probably a part of the Great Rift.

This is without any doubt, a special part of the Milky Way and one can spend literally hours just scanning the sky and enjoying the views.

The constellation contains many objects: clusters, nebulae and of course double stars, and to catalog them all could easily fill a small book, so we shall look at just the best it has to offer. Due to the dust and gas present, galaxies are few and none are bright. Let's starts as usual by looking at some stars.

Without a doubt the finest double star in the constellation, and perhaps the finest in the northern sky, is **Albireo**, or **Beta (β) Cygni**, a golden-yellow primary and lovely blue secondary against the backdrop of the myriad fainter stars of the Milky Way. It is easy to locate at the foot of the Northern Cross. The colors can be made to appear even more spectacular if you slightly defocus the images. There is evidence that the brighter star is itself a double and it can be observed with telescopes of aperture 50 cm and larger. Wonderful! Another fine, but difficult, double is **Delta (δ) Cygni**. Contrasting reports of this system's colors abound: a blue-white, pale-yellow or greenish-white primary, and a blue-white or bluish secondary. It is a difficult object for southern observers because of its northerly declination. However, it does make a good test for telescopes of 10–15 cm, but exceptional seeing is needed.

An easy double is **17 Cygni** that is set amongst a wonderful star field. The brightest star is a lovely yellow color which contrasts nicely with the fainter orange companion, located to its northeast. In addition, some 27 arcminutes to the southwest of the pair is a very nice faint orange pair of stars, **Σ (Struve) 2576**, both at 8th magnitude. Some other fine doubles are **49 Cygni** and **OΣ 437**. The former consists of a bright yellow primary and a white or bluish secondary that can easily be resolved in an 8 cm telescope. The latter is a lovely orange-yellow pair of stars that can also be split with a small telescope.

Something of a test is **B 677**. This consists of a bright orange-yellow star located in a most wonderful star field and would be worth observing for that alone, but there is a faint star to the southeast that will need careful observation in order to be seen.

A nice triple star system is **OΣ 390**, again set in a star-filled field of view. It consists of a bright pale-yellow star that has a companion to its northwest. This should be easily split with a 8 cm telescope, but a 20 cm aperture will be needed to see the third member, which is a fainter star to the south.

There are a couple of variable stars we can also look at. The first is **RS Cygni** that has the distinction of having a lovely color when it is near or at its maximum brightness. This is a red giant star with a persistent periodicity, class SRA, and has a period of 417.39 days, with a magnitude range of 6.5–9.5. It is a strange star where the light curve can vary appreciably, with the maxima sometimes doubling. However, what makes it so special is its deep red color. Try comparing it with close-by **Theta (θ) Cygni**, a white star. The color contrast between them often seems to make Theta take on a distinct bluish tint. Another variable is **Chi (χ) Cygni**. What makes this long-period variable so special is the large range of magnitudes it exhibits. Perfect for binocular observation, the orange star can be as bright as 3rd magnitude, and so visible to the naked eye, and then over about 200 days, it fade down to 14th magnitude, where it begins the cycle over again. It lies about a quarter of the way from **Eta (η) Cygni** to Albireo.

We cannot discuss Cygnus without mentioning its brightest star, **Deneb, Alpha (α) Cygni**. The twentieth-brightest star in the sky is very familiar to observers in the northern hemisphere. This pale-blue supergiant has recently been recognized as the prototype of a class of nonradially pulsating variable stars. Although the magnitude change is very small, the timescale is from days to weeks. It is believed that the luminosity of Deneb is some 60,000 times that of the Sun, with a diameter 60 times greater. It is a very nice pale-blue color and is the faintest star of the Summer Triangle, the other members being Vega and Altair.

Our penultimate star is the famous **61 Cygni**. This star is best seen with binoculars (but is sometimes a challenge if conditions are poor), which seem to emphasize the vibrant colors of

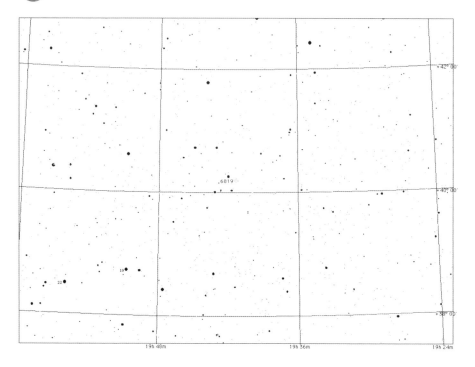

these stars, both orange-red and K-type. It is famous for being the first star to have its distance measured by the technique of parallax, when German astronomer Friedrich Bessel determined its distance to be 10.3 light years; modern measurements give a figure of 11.36, thus making it the fourteenth-nearest star to us. It also has an unseen third component, which has the mass of eight Jupiters. The system also has a very large proper motion.

Now for something quite odd, but very fascinating. The star **16 Cygni B** is a nondescript star to the observer, but it does possibly have one very important attribute. It may have a planet orbiting it! The star is a visual binary, and the companion, **16 Cygni A,** is about 700 AU away. The planet also has a very large eccentricity,[23] value 0.6, which is causing some concern among astronomers, as they cannot explain it! This may imply the presence of orbiting planets. Try observing the star and think that it may have a retinue of planets, and who knows what else?

Now for some open clusters. First, though, let me say that there are a lot of cataloged star clusters in Cygnus. However, many are faint and so almost impossible to tell apart from the magnificent background of the Milky Way. And, when you come to think about it, who needs clusters here when the constellation is very much just like one gigantic and spectacular cluster anyway! But we need to be scientific and precise, so our first is **NGC 6819 (Collinder 403)** (see Star Chart 2.52). This is a rich and distant open cluster with an integrated magnitude of 7.3, and is located within and contrasting with the Milky Way (see Figure 2.42). It contains many 11th magnitude stars, and thus is an observing challenge. The cluster is very old at over 3 billion years.

---

[23] Eccentricity is just a measure of how circular the orbit is.

Figure 2.42. NGC 6819 (Harald Strauss, AAS Gahberg).

Then there is **NGC 6871** (**Collinder 413**) (see Star Chart 2.53). This is a nice cluster that is easily seen in small telescopes. It does, however, appear as an enhancement of the background Milky Way (see Figure 2.43). Binoculars will show several stars of 7th to 9th magnitude surrounded by fainter members in an area about half a degree across. It also includes the nice orange star **27 Cygni**.

A cluster that is often overlooked by nearly everyone is **Roslund 5.** It is a largish but obscure object that can be found about halfway between **Eta Cygni** and **39 Cygni**, consisting of around 20 stars from 7th to 10th magnitude. Although it is best seen through

Star Chart 2.53. NGC 6871.

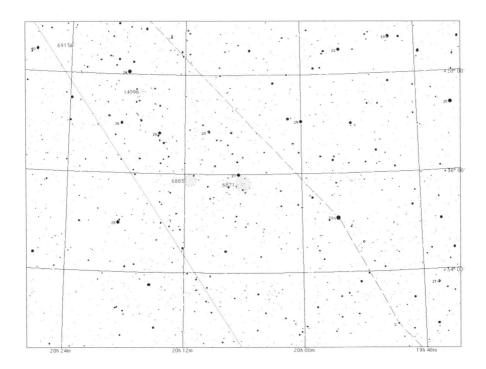

Figure 2.43. NGC 6871 (Space Telescope Science Institute, AAO, UK–PPARC, ROE, National Geographic Society, and California Institute of Technology).

binoculars, due to the plethora of background stars it is difficult to discern where the cluster begins and ends.

No constellation would be complete without its Messier objects, and so we have **Messier 29 (NGC 6913)**. This is a very small cluster and one of only two Messier objects in Cygnus. It contains only about a dozen stars visible with small instruments, and even then benefits from a low magnification (see Figure 2.44). However, studies show that it contains many more bright B0-type giant stars, which are obscured by dust. Without this, the cluster would be a very spectacular object.

The other Messier object is **Messier 39 (NGC 7092)**. This is a nice cluster in binoculars, and lies at a distance of 840 light years. There are about two dozen stars visible, ranging from 7th to 9th magnitude. What makes this cluster so distinctive is the lovely color of the stars – steely blue – and the fact that it is nearly perfectly symmetrical, having a triangular shape (see Figure 2.45). There is also a nice double star at the center of the cluster. Under good conditions it can be glimpsed with the naked eye.

Figure 2.44. Messier 29 (Harald Strauss, AAS Gahberg).

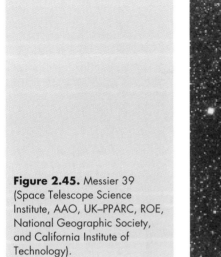

Figure 2.45. Messier 39 (Space Telescope Science Institute, AAO, UK–PPARC, ROE, National Geographic Society, and California Institute of Technology).

Our final open cluster is another one of those that are often overlooked because it is difficult to distinguish its members from the background. It is called **Ruprecht 173** and is a large bright object. At nearly 1° across it has about 30 members of 9th magnitude and brighter that are best seen in low-power binoculars. It lies about one-quarter of the way

Figure 2.46. NGC 6894 (Space Telescope Science Institute, AAO, UK–PPARC, ROE, National Geographic Society, and California Institute of Technology).

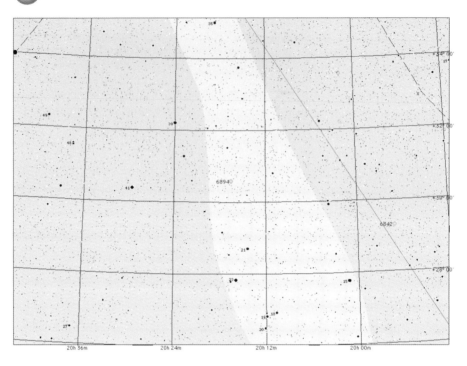

Star Chart 2.54 (*above*). NGC 6894.
Star Chart 2.55 (*below*). NGC 7026; NGC 7027; NGC 7048.

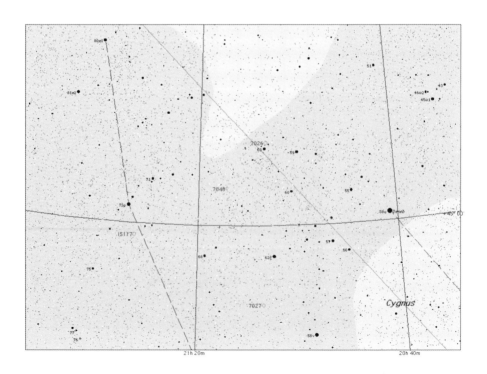

Figure 2.47. NGC 7026 (Klaus Eder and Georg Emrich, AAS Gahberg).

from **Epsilon (ε) Cygni** to **Gamma Cygni**. There is also a nice Cepheid variable star within the cluster, **X Cygni**, which has a 16-day period varying from 6th to 7th magnitude.

There are a few planetary nebulae in Cygnus, although most are faint and small. But there are some that are suitable for us. The first is **NGC 6894**. First discovered in 1784, this is a faint circular nebula about 45 arcseconds across (see Figure 2.46). It is admittedly a difficult object, but should be visible in a telescope of 30 cm aperture. However, you will need care and patience to find it (see Star Chart 2.54).

Three planetaries that can be seen in telescopes of about 20 cm aperture are **NGC 7026**, **NGC 7027** and **NGC 7048** (see Star Chart 2.55). The first is a bright blue-green colored object, some 15 arcseconds in diameter, that exhibits a brightening towards its center (see Figure 2.47). The second object can be seen as a bluish elliptical nebula, some 10 × 5 arcseconds (see Figure 2.48). The last nebula can be seen as a faint 1 arcminute disk (see Figure 2.49). It must be admitted that larger telescopes as well as the use of an [OIII] filter will give better views with these planetaries, showing far more detail – and in the first object an occasional glimpse of a central star – but under superb conditions they may be glimpsed with telescopes as small as 10 cm, appearing as out-of-focus stars.

Our penultimate planetary is perhaps the most famous in Cygnus: **NGC 6826** (**Caldwell 15**; see Star Chart 2.56). Also known as the **Blinking Planetary** this will be a difficult planetary nebula to locate, but will be well worth the effort. It shines at 9th magnitude, and using averted vision will appear as a tiny glow in small telescopes (see Figure 2.50). The

Figure 2.48. NGC 7027 (Robert Schulz, AAS Gahberg).

Figure 2.49. NGC 7048 (Klaus Eder and Georg Emrich, AAS Gahberg).

so-called blinking effect is due solely to the physiological structure of the eye. If you stare at the central star long enough, the planetary nebula will fade from view. At this point should you move the eye away from the star, and the planetary nebula will "blink" back into view at the periphery of your vision. Visually, it is a nice blue-green disk, which will take high magnification well. Although not visible in amateur telescopes, the planetary nebula is made up of two components – an inner region consisting of a bright shell and

Star Chart 2.56. NGC 6826.

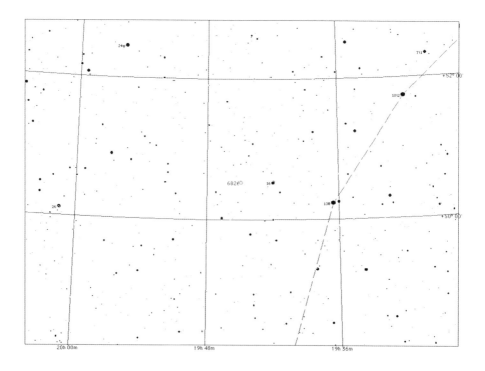

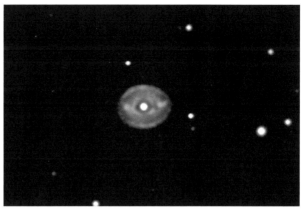

Figure 2.50. NGC 6826 (Robert Schulz, AAS Gahberg).

two ansae, and a halo that is delicate in structure with a bright shell. Large binoculars will show the central star shining at 11th magnitude.

Our final planetary nebula is **PK 64 + 5.1** and I only include it because of its central star. This is a very small planetary nebula, which even with apertures of 20 cm and greater will require a high magnification. What makes it even more difficult to locate is the multitude of stars in the background. However, a pointer to the planetary nebula is the star responsible for it – **Campbell's Hydrogen Star**, which has a lovely orange color.

Now let us look at some of the most famous objects in Cygnus – emission nebulae. Probably the most famous object is **NGC 7000 (Caldwell 20)**, also known as the **North America Nebula** (see Star Chart 2.57). This is a famous emission nebula, visible on dark nights to the naked eye (see Figure 2.51). Located just 3° east of Deneb, it is magnificent in binoculars, melding as it does into the stunning star fields of Cygnus. Providing you know where, and what to look for, the nebula is visible to the naked eye. However, it is rather low for southern observers. With small- and large-aperture telescopes details within the nebula become visible, though several amateurs have reported that increasing aperture decreases the nebula's impact. The dark nebula lying between it, known as **Lynds 935**, and the Pelican Nebula (see below) is responsible for its characteristic shape. Until recently, Deneb was thought to be the star responsible for providing the energy to make the nebula glow, but recent research points to several unseen stars being the power sources.

The **Pelican Nebula**, **IC 5067/70**, lies close to the North American Nebula (see the entry above) and has been reported to be visible to the naked eye. It is easily glimpsed in binoculars as a triangular faint hazy patch of light and can be seen best with averted vision, and the use of light filters. A good test to see whether the conditions are right to observe the nebula is to determine whether or not the North American Nebula is visible to the naked eye. If it is, then you should be able to see the Pelican Nebula with large binoculars and small telescopes.

One of the most photographed objects in Cygnus is **NGC 6888 (Caldwell 27)**, also known as the **Crescent Nebula** (see Star Chart 2.58). Although visible in small telescopes, a dark location and a light filter will make its detection much easier (see Figure 2.52). With good conditions, the emission nebula will live up to its name, having an oval shape with a gap in the ring on its southeastern side. The nebula is known as a **stellar wind bubble**,[24] and is the result of a fast-moving stellar wind from a Wolf–Rayet star which is sweeping up all the material that it had previously ejected during its red giant stage. It can be seen from urban locations but you will need a large aperture telescope and a [OIII] filter. Surprisingly, there are also reports that under very dark skies, and in areas with no light pollution, it can be

---

[24] A more recent name is "wind-blown Wolf–Rayet ring nebula".

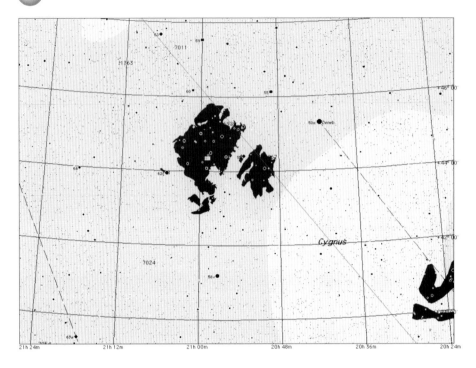

Star Chart 2.57. North America Nebula; Pelican Nebula.

glimpsed with binoculars. It lies some 2.75° southwest of Gamma Cygni. In fact, the whole area surrounding Gamma Cygni is immersed in nebulosity and is an easy target for astrophotographers and CCD imagers.

Another nebula is **IC 5146** (**Caldwell 19**), also known as the **Cocoon Nebula** (see Star Chart 2.59). This is a very difficult nebula to find and observe because it has a low surface brightness and appears as nothing more than a hazy amorphous glow surrounding a couple of 9th magnitude stars (see Figure 2.53).

The dark nebula **Barnard 168** (which the Cocoon lies at the end of) is surprisingly easy to find, and thus can act as a pointer to the more elusive emission nebula. It is an easily distinguished dark nebula that extends from the western edge of IC 5146. In binoculars its large size, some 10 × 100 arcminutes, can be easily spotted set against the innumerable background stars. The whole area comprising both IC 5146 and Barnard 168 is a vast stellar nursery and recent infrared research indicates the presence of many new and proto-stars within the nebula itself.

A strange object that has been the source of some debate is **CRL 2688** (**PK 80–6.1**), which is also known as the **Egg Nebula**. It has been called a bipolar reflection nebula and a plane-tary nebula. It is a very difficult object and will need the best observing conditions. Exhibiting an elongated shape, about 20 × 10 arcseconds, it can be seen in a 30 cm tele-scope. A novel idea is to rotate some polarized film (or glasses) in front of the eyepiece; this will dim the light by 1 magnitude, and indicates that the light is highly polarized. It is classed as one of the rare protoplanetary nebulae.

Dark nebulae abound in Cygnus, and are part of the celestial panorama that the constel-lation offers (see Figure 2.54). Of these, the following are the easiest to see. Visible in

Figure 2.51 North America Nebula; Pelican Nebula (Matt BenDaniel, http://starmatt.com).

Figure 2.52. NGC 6888 (Michael Karrer, AAS Gahberg),

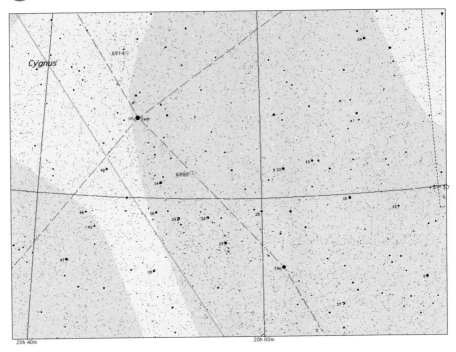

Star Chart 2.58 (*above*). NGC 6888.
Star Chart 2.59 (*below*). IC 5146.

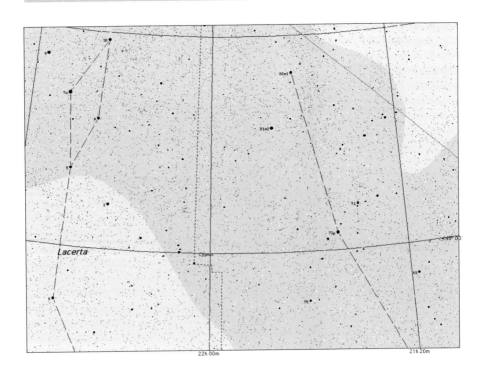

Figure 2.53. Cocoon Nebula and Barnard 168 (Matt BenDaniel, http://starmatt.com).

Figure 2.54. Dark nebulae in central Cygnus (Matt BenDaniel, http://starmatt.com).

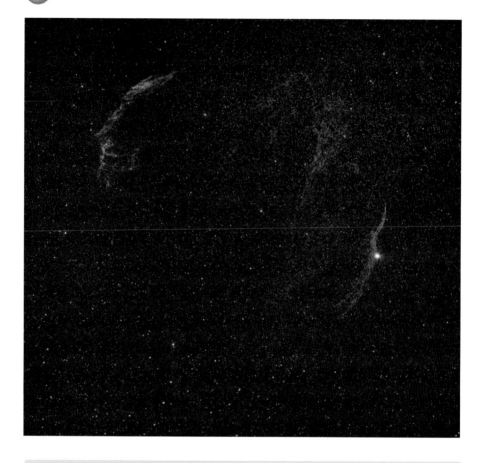

Figure 2.55. Veil Nebula (Matt BenDaniel, http://starmatt.com).

binoculars, **Barnard 352** is part of the much more famous North American Nebula, though this dark part is located to the north. It is a well-defined triangular dark nebula. A smaller area is **Barnard 343**. It can easily be seen as a "hole" in the background Milky Way as an oval dark nebula, which although glimpsed in binoculars is at its best in telescopes. Visible in binoculars is **Barnard 145** as a triangular dust cloud that stands out well against the impressive star field. As it is not completely opaque to starlight, several faint stars can be seen shining through it. Finally there is **Lynds 906**, also known as the **Northern Coalsack.**

This is probably the largest dark nebulosity of the northern sky. It is an immense region, easily visible on clear moonless nights just south of Deneb. It lies just at the northern boundary of the Great Rift, a collection of several dark nebulae which bisects the Milky Way. The Rift is of course part of a spiral arm of the Galaxy, that features prominently on photographs of other galaxies such as **NGC 891** in Andromeda. Our final dark nebula is **Harrington 10**, which is a small patch of nebulosity lying about 7° northeast of Deneb. It can be seen with the naked eye on clear nights as a dark line perpendicular to the Milky Way. Using binoculars will show some faint structure along its perimeter that are often portrayed in deep images.

Our final object is a magnificent supernova remnant (see Figure 2.55). For our purposes we can say it is in three parts. The first, **NGC 6960 (Caldwell 34)**, is also known as the **Veil**

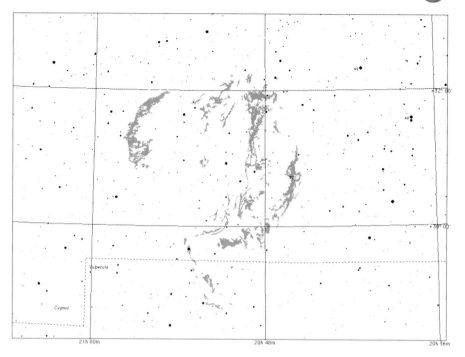

Star Chart 2.60. Veil Nebula.

Nebula (Western Section). This is the western portion of the **Great Cygnus Loop**, which is the remnant of a supernova that occurred about 30,000 years ago. It is easy to locate because it is close to the star **52 Cygni**, though the glare from this star makes it difficult to see. The star itself is a close double of yellow and orange stars. Dark skies are needed and a light filter makes a vast difference. Positioning the telescope so that 52 Cygni is out of the field of view also helps. The nebulosity we observe is the result of the shockwave from the supernova explosion impacting on the much denser interstellar medium. So far the actual remains of the star have yet to be detected.

The second part is **NGC 6992** (**Caldwell 33**), which is also known as the **Veil Nebula (Eastern Section)** (see Star Chart 2.60). A spectacular object when viewed under good conditions and brighter than NGC 6960, it is the only part of the Loop that can be seen in binoculars, and has been described as looking like a fish-hook. It takes large aperture and high magnification well, and 40 cm telescopes will show the southern knot. Using such a telescope, it becomes apparent why the nebula has been named the **Filamentary Nebula**, as lacy and delicate strands will be seen. However, there is a downside: it is notoriously difficult to find. Patience, clear skies and a good star atlas will help. This is a showpiece of the summer sky (when you have finally found it). The final part is **NGC 6974–79**, and is also known as the **Veil Nebula (Central Section)**.

This part of the Great Cygnus Loop is difficult to see, but the use of light filters makes it easier to locate and observe. It appears as a triangular hazy patch of light. A very transparent sky is needed to glimpse this, as indeed it is for all the sections of this wonderful object. I recall that the best view I had of the nebula was after a rainstorm when the atmosphere was very transparent and dark.

2.10 Lyra

One of the smallest constellations in the sky, Lyra is also probably one of the most recognized. It has a very distinctive shape, reminiscent of a parallelogram, and is always a highlight of the summer sky for northern observers (see Star Chart 2.61). It lies on the fringes of the Milky Way, and has many delightful objects that we can look at. It transits in early July. For southern observers it is low down, but will still allow some decent observation to be made.

We start as usual with stars, but let me state straightaway that the Milky Way does not encompass two of Lyra's most famous members, **Vega** and **Epsilon (ε) Lyrae**. It is a sad twist of fate, but there you go. But there are a few double and variable stars we can look at, so let's begin with those.

A nice but difficult double is **Beta (β) Lyrae**. This pair of white stars is a challenging double for binoculars. β^1 is also an eclipsing binary with two unequal minima of 3.8 and 4.3 separated by a maximum of 3.4 over a period of 12.91 days. A fascinating situation occurs owing to the gravitational effects of the components of β^1. The stars are distorted from their spherical shapes into ellipsoids. The star is also a multiple system, with a 7.8 magnitude star, a couple of 9.9 magnitude stars and a 13th magnitude star.

Another famous multiple system is Lyra's other "double-double", Σ **(Struve) 2470** and Σ **(Struve) 2474**. These are a pair of well-separated stars that can be seen in the same field of view about 11 arcminutes apart and are very similar in all respects apart from their color. The Σ 2470 stars are white and bluish white, whereas the Σ 2474 stars are both pale yellow.

Star Chart 2.61. Lyra.

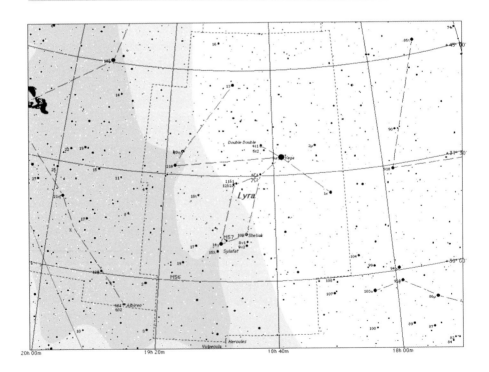

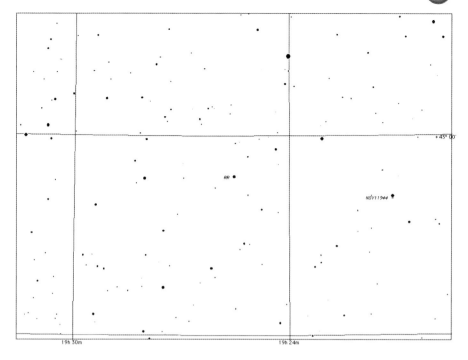

Star Chart 2.62. RR Lyrae.

Both of the systems are an easy object to observe for small telescopes. A lovely little system is **Eta (η) Lyrae,** which consists of a nice pair of stars along with three other smaller pairs, all within about 7 arcminutes of each other. The brighter pair are white, but for southern observers who will be observing them at a low altitude, the colors may appear pale yellowish.

One of the most famous variable stars in Lyra is **RR Lyrae**, the prototype of the cluster of pulsating variable stars (see Star Chart 2.62). These are similar to Cepheid variable stars but have shorter periods and lower luminosities. There are no naked-eye members of this class of variable, and RR Lyrae is the brightest member. There is a very rapid rise to maximum, with the light of the star doubling in less than 30 minutes, with a slower falling in magnitude. From an observational viewpoint, it is a nice white star, although detailed measurements have shown that it does become bluer as it increases in brightness. There is some considerable debate as to the changes in spectral type that accompany the variability. One source quotes A8–F7, while another A2–F1. Take your pick. There is also some indications that there is another variability period along with the shorter one, which has a period of about 41 days. It varies in magnitude from 7.06 to 8.12.

There are a couple of open clusters that can be observed here, including the lesser-known **Stephenson 1.** This is seen as a large but loose group of stars in binoculars and small telescopes that also contains its brightest members, **Delta^1 (δ^1)** and **Delta^2 (δ^2) Lyrae** (see Star Chart 2.63). These two stars are believed to be real members of the cluster and are separated by about 10 arcminutes and surrounded by 12 fainter cluster members, lying predominantly to the west. There is a delightful color contrast between the nice orange of the 4.5 magnitude Delta^2 Lyrae and the blue-white of the 5.5 magnitude Delta^1 Lyrae. Using

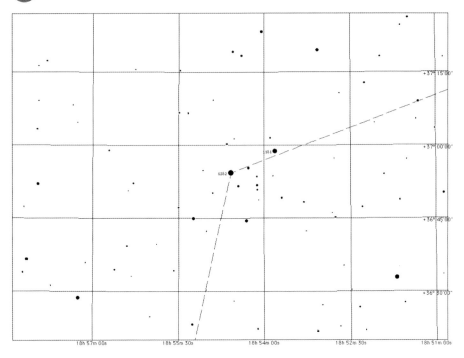

Star Chart 2.63. Stephenson 1.

a larger aperture will show many more members. It is one of the nearest open clusters at 800 light years.

Another cluster is **NGC 6791** (see Star Chart 2.64). This is a rich cluster of faint stars that contains many faint 11th magnitude stars and so poses an observing challenge (see Figure 2.56). With small telescopes a couple of dozen stars can be resolved against a hazy background, but larger apertures will of course reveal many more, maybe several hundred, cluster stars.

As is to be expected, there are the requisite Messier objects. The first is the globular cluster **Messier 56** (**NGC 6779**). With a magnitude of 8.3, and a diameter of around 7 arcminutes, the cluster is situated in a rich star field and in small instruments will appear as a hazy patch with a brighter core (see Figure 2.57). It has often been likened to a comet in its appearance. Resolution of the cluster will need at least a 20 cm aperture telescope, and increasing magnification will show further detail. See Star Chart 2.61.

The other Messier object is of course the finest object in Lyra, the wonderful **Ring Nebula, Messier 57** (**NGC 6720**). This is probably the most famous of all planetary nebulae, and surprisingly – and pleasantly – visible in binoculars, shining with a magnitude of 8.8 and over 50 arcminutes in diameter. However, it will not be resolved into the famous "smoke-ring" shape seen so often in color photographs; it will, rather, resemble an out-of-focus star. It is just resolved in telescopes of about 10 cm aperture, and at 20 cm the classic smoke-ring shape becomes apparent (see Figure 2.58). At high magnification (and larger aperture), the Ring Nebula is truly spectacular. The inner region will be seen to be faintly hazy, but large aperture and perfect conditions will be needed to see the central star. Does the planetary nebula appear perfectly circular, or is it slightly oval? See Star Chart 2.61.

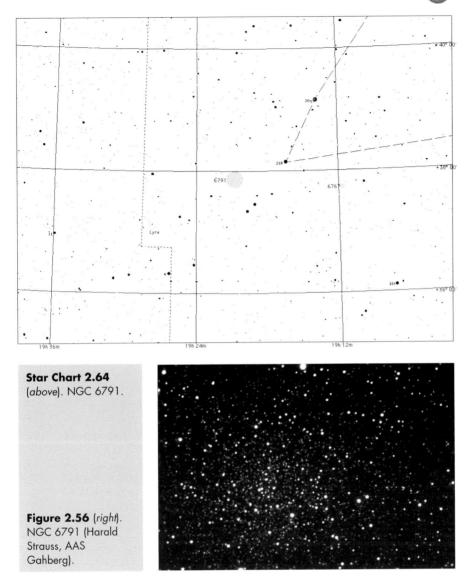

Star Chart 2.64
(*above*). NGC 6791.

Figure 2.56 (*right*).
NGC 6791 (Harald
Strauss, AAS
Gahberg).

There are several galaxies in Lyra, but sadly they are very faint and need apertures of about 40 cm and greater, so we shall not discuss them.

2.11 Lacerta

Another little constellation that is passed over by many amateurs is **Lacerta**. It's not surprising really as it doesn't have a lot to offer the small telescope owner, and for southern observers it is very low in the sky, so tends to be forgotten. That is a shame as it does hold

Figure 2.57. Messier 56 (Harald Strauss, AAS Gahberg).

a few nice objects. The Milky Way actually covers the entire constellation even though on many star atlases it will only show the northern half being covered (see Star Chart 2.65). I shall, as usual, take the more liberal approach. It transits at the end of August.

Lacerta has a fair number of nice double stars that show a color contrast, and of these Σ (**Struve**) **2876**, Σ (**Struve**) **2894 and** Σ (**Struve**) **2940** are particularly good. The first system has a nice contrast of white and blue stars that are easily split in telescopes of 10 cm or more. The second double, again easily resolved in small telescopes, is a nice yellowish and blue pair of stars, although southern observers may see the pair as yellow and red due to the low altitude of the stars. The third system is a close 6 arcsecond pair of yellowish and white stars. In larger telescopes another double can be seen to the northeast at magnitudes 12.1 and 12.9 separated by 5 arcseconds. Another fine double is Σ (**Struve**) **2942**, which is a fine orange and white system, but sometimes the fainter star may appear as green due to the color contrast. There is a faint third companion with magnitude 11.5 that may not be visible to southern observers. One star system that could easily be called a

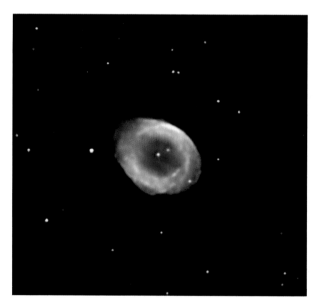

Figure 2.58. Messier 57 (Michael Karrer, AAS Gahberg).

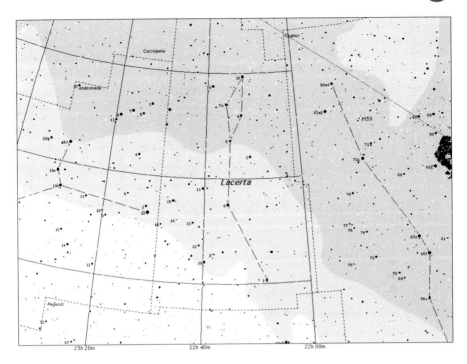

Star Chart 2.65. Lacerta.

small cluster instead of a multiple star is **8 Lacertae** (Σ **2922**). In telescopes of about 20 cm or so there should appear to be four bluish-white and white stars all within 84 arcseconds of each other. A fifth member lies about 5.5 arcminutes away to the southwest and there are also several fainter 13th and 14th magnitude stars nearby.

For binoculars and small telescopes, there are only two clusters that are easily seen, and these are NGC 7209 and NGC 7243. The former, **NGC 7209 (Herschel 53)**, shining with an integrated magnitude of 7.7, is set amongst a lovely star-strewn region of the Milky Way and is a large cluster with about 75 or more 10th magnitude members (see Star Chart 2.66). In small telescopes or binoculars this will appear as a hazy glow upon which will be a few 9th magnitude stars (see Figure 2.59). In larger telescopes several arcs and chains of stars can be observed.

The latter cluster, **NGC 7243 (Caldwell 16)**, is fairly bright at magnitude 6.4, and is a large, irregular cluster that although amongst myriad stars of the Milky Way nevertheless stands out quite well (see Star Chart 2.66). Several of the stars are visible in binoculars, but the remainder blur into the background star field (see Figure 2.60). A nice object in an otherwise empty part of the sky – if you overlook the fact that it is located within the Milky Way!

A cluster that may need a larger aperture of, say, 25 cm in order to be really appreciated is **IC 1434 (Collinder 445)** (see Star Chart 2.66). Although it has a magnitude of about 9.0, this can be a somewhat difficult cluster to locate and observe (see Figure 2.61). Located within the Milky Way, this is a large but irregular cluster of over 70 stars of 10th magnitude and fainter and it may be wise to try using a high magnification of, say, 150 to 200×,

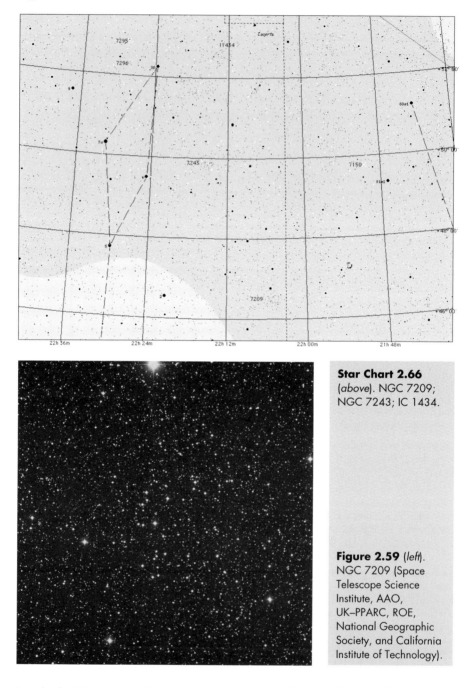

Star Chart 2.66 (*above*). NGC 7209; NGC 7243; IC 1434.

Figure 2.59 (*left*). NGC 7209 (Space Telescope Science Institute, AAO, UK–PPARC, ROE, National Geographic Society, and California Institute of Technology).

in order for it to be seen. Also try to use averted vision. These two factors will almost certainly improve this cluster.

The remainder of the open clusters in Lacerta, and all of the galaxies, of which there are several, will need telescopes of 30 cm aperture and larger to be appreciated, so with

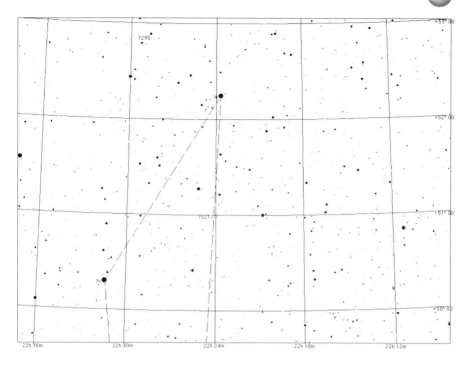

Star Chart 2.67. IC 5217.

that in mind, I will leave them to large-aperture telescope owners to seek out for themselves.

There is one planetary nebula that may be of interest: **IC 5217**. It is a tiny object, only 8×6 arcseconds, and so in most cases will appear star-like (see Star Chart 2.67). It has a slightly bluish color which is of course more readily seen in large-aperture telescopes as well as those equipped with an [OIII] filter. Although there isn't much to see, it is worthwhile seeking out and crossing off your list of "things to observe".

Perhaps I ought to close this section and chapter by mentioning what is probably Lacerta's most famous resident, but one that is nigh on impossible to observe with even the largest amateur telescopes. This is **BL Lacertae**, the prototype of a class of active galaxy that is characterized by a lack of emission lines in their spectrum and by rapid and large magnitude variations. With BL Lac, as it is known, the magnitude can vary between 14th and 17th magnitude. The vast energy output is believed to be caused by material – either dust, gas and stars – falling into a massive black hole. A very exotic and strange object indeed but from an amateur astronomer's point of view, forever invisible.

The following constellations are also visible during these months at different times throughout the night. Remember that they may be low down and so diminished by the effects of the atmosphere. Also, you may have to observe them either earlier than midnight, or some considerable time after midnight, in order to view them.

Figure 2.60. NGC 7243 (Space Telescope Science Institute, AAO, UK–PPARC, ROE, National Geographic Society, and California Institute of Technology).

Northern Hemisphere

Andromeda, Aquila, Camelopardalis, Cassiopeia, Cepheus, Cygnus, Delphinus, Gemini, Hercules, Lacerta, Libra, Lupus, Lyra, Orion, Ophiuchus, Perseus, Sagitta, Sagittarius, Scutum, Scorpius, Serpens Cauda, Taurus, Vulpecula.

Figure 2.61. IC 1434 (Space Telescope Science Institute, AAO, UK–PPARC, ROE, National Geographic Society, and California Institute of Technology).

Southern Hemisphere

Andromeda, Antila, Apus, Ara, Camelopardalis, Carina, Cassiopeia, Centaurus, Cepheus, Chamaeleon, Circinus, Corona Australis, Crux, Libra, Lupus, Musca, Norma, Octans, Ophiuchus, Pavo, Scorpius, Telescopium, Triangulum Australe, Vela, Volans.

Objects in Sagittarius

Stars

Designation	Alternate name	Vis. mag	RA	Dec.	Description
Sagittarius A*			$17^h 45.6^m$	−20° 00'	Center of Galaxy
h 5003	Herschel 5003	5.2, 6.9	$17^h 59.1^m$	−30° 15'	PA 105°; Sep. 5.5"[1]
21 Sagittarii	Jc 6	4.9, 7.4	$18^h 25.3^m$	−20° 32'	PA 289°; Sep. 1.8"
HN 119		5.6, 8.6	$19^h 29.9^m$	−26° 59'	PA 142°; Sep. 7.8"
AQ Sagittarii		6.6–7.6	$19^h 34.3^m$	−16° 22'	Carbon star
54 Sagittarii	h 599	5.4, 11.9, 8.9	$19^h 40.7^m$	−16° 18'	PA 274°; Sep. 38.0"AB/ PA 42°; Sep. 45.6"AC
Eta (η) Sagittarii	β 760	3.2, 7.8	$18^h 17.6^m$	−36° 46'	PA 105°; Sep. 3.6"

Deep-Sky Objects

Designation	Alternate name	Vis. mag	RA	Dec.	Description
NGC 6494	Messier 23	5.5[2]	$17^h 56.8^m$	−19° 01'	Open cluster
NGC 6520	Herschel 7	7.6p[3]	$18^h 03.4^m$	−27° 54'	Open cluster
NGC 6531	Messier 21	5.9p	$18^h 04.6^m$	−22° 30'	Open cluster
Messier 24	Small Sagittarius Star Cloud	4.6	$18^h 16.5^m$	−18° 50'	Star cluster
NGC 6603		11.1p	$18^h 18.4^m$	−18° 25'	Open cluster
NGC 6613	Messier 18	6.9	$18^h 19.9^m$	−17° 08'	Open cluster
IC 4725	Messier 25	4.6	$18^h 31.6^m$	−19° 15'	Open cluster
NGC 6645	Herschel 23	8.5p	$18^h 32.6^m$	−16° 54'	Open cluster
NGC 6440	Herschel 150	9.1	$17^h 48.9^m$	−20° 22'	Globular cluster
NGC 6522	Herschel 49	8.6	$18^h 03.6^m$	−30° 02'	Globular cluster
NGC 6528	Herschel 200	9.5	$18^h 04.8^m$	−30° 03'	Globular cluster
NGC 6544	Herschel 197	8.1	$18^h 07.3^m$	−25° 00'	Globular cluster
NGC 6553	Herschel 12	8.1	$18^h 09.3^m$	−25° 54'	Globular cluster
NGC 6626	Messier 28	6.8	$18^h 24.5^m$	−24° 52'	Globular cluster
NGC 6637	Messier 69	7.6	$18^h 31.4^m$	−32° 21'	Globular cluster
NGC 6656	Messier 22	5.1	$18^h 36.4^m$	−23° 54'	Globular cluster

[1] Bear in mind that the position angle may change over a short period of time and so the values given here may be different from what you observe now.

[2] The magnitude given for deep-sky objects is the integrated magnitude.

[3] The subscript indicates the photographic magnitude.

Name	Alternative name	Magnitude	RA	Dec	Type
Palomar 8		10.9	$18^h41.5^m$	$-19° 49'$	Globular cluster
NGC 6681	Messier 70	8.0	$18^h42.2^m$	$-32° 18'$	Globular cluster
NGC 6715	Messier 54	7.6	$18^h55.1^m$	$-30° 29'$	Globular cluster
Palomar 9	NGC 6717	9.2	$18^h55.1^m$	$-22° 42'$	Globular cluster
NGC 6723		7.2	$18^h59.6^m$	$-36° 38'$	Globular cluster
NGC 6809	Messier 55	6.4	$19^h40.0^m$	$-30° 58'$	Globular cluster
NGC 6476		–	$17^h53.8^m$	$-29° 00'$	Star field
NGC 6514	Messier 20/ Trifid Nebula	–	$18^h02.3^m$	$-23° 02'$	Emission & reflection nebula
NGC 6523	Messier 8/ Lagoon Nebula	–	$18^h03.8^m$	$-24° 23'$	Emission nebula
NGC 6530		4.6p	$18^h04.8^m$	$-24° 20'$	Open cluster
NGC 6618	Messier 17/ Omega Nebula	–	$18^h20.8^m$	$-16° 11'$	Emission nebula
NGC 6589 6590		–	$18^h16.9^m$	$-19° 47'$	Reflection nebula
Barnard 289		–	$17^h56.4^m$	$-28° 55'$	Dark nebula
Barnard 87	Parrot's Head Nebula	–	$18^h04.3^m$	$-32° 30'$	Dark nebula
Barnard 92		–	$18^h15.5^m$	$-18° 14'$	Dark nebula
Barnard 86	Ink Spot Nebula	–	$18^h13.0^m$	$-27° 53'$	Dark nebula
NGC 6445		11.2	$17^h49.2^m$	$-20° 01'$	Planetary nebula
NGC 6565	PK8+3.1	11.6	$18^h11.9^m$	$-28° 11'$	Planetary nebula
NGC 6563	PK3–4.5	11.0	$18^h12.0^m$	$-33° 52'$	Planetary nebula
NGC 6567	PK358–7.1	11.0	$18^h13.7^m$	$-19° 05'$	Planetary nebula
NGC 6629	PK11–0.2	11.3	$18^h25.7^m$	$-23° 12'$	Planetary nebula
NGC 6818	PK9–5.1 Little Gem	9.3	$19^h44.0^m$	$-14° 09'$	Planetary nebula
NGC 6822	Caldwell 57/ Barnard's Galaxy	8.8	$19^h44.9^m$	$-14° 48'$	Galaxy

Objects in Serpens Cauda

Stars

Designation	Alternate name	Vis. mag	RA	Dec.	Description
5 Serpentis		5.1, 9.7	$15^h 19.2^m$	+01° 46'	PA 36°; Sep. 11"
Nu (ν) Serpentis		4.3, 8.3	$17^h 20.8^m$	−12° 51'	PA 28°; Sep. 46.3"
Σ 2303	Struve 2303	6.6, 9.1	$18^h 20.1^m$	−07° 59'	PA 236°; Sep. 2.1"
AC (Alvin Clark) 11		6.8, 7.0	$18^h 24.9^m$	−01° 35'	PA 355°; Sep. 0.8"
59 Serpentis	Struve 22316	5.3, 7.6	$18^h 27.2^m$	+00° 12'	PA 318°; Sep. 3.8"
Theta (θ) Serpentis		4.6, 5.0	$18^h 56.2^m$	+04° 12'	PA 104°; Sep. 22.3"

Deep-Sky Objects

Designation	Alternate name	Vis. mag	RA	Dec.	Description
IC 4756		4.6	$18^h 39.0^m$	+05° 27'	Open cluster
NGC 6611	Messier 16	6.0	$18^h 18.8^m$	−13° 47'	Open cluster
IC 4703	Eagle Nebula	–	$18^h 18.6^m$	−13° 58'	Emission nebula
NGC 6535		10.6	$18^h 03.9^m$	−00° 18'	Globular cluster
NGC 6539		9.8	$18^h 04.8^m$	−07° 35'	Globular cluster

Objects in Scutum

Stars

Designation	Alternate name	Vis. mag	RA	Dec.	Description
R Scuti		4.9–8.2	$18^h 47.5^m$	−05° 42'	Variable star
Delta (δ) Scuti		$4.7_v,^4$ 9.2	$18^h 42.3^m$	−09° 03'	Variable star
Σ 2306	Struve 2306	7.9, 8.6	$18^h 22.2^m$	−15° 05'	PA 221°; Sep. 10.2"
Σ 2373	Struve 2373	7.2, 8.2	$18^h 45.9^m$	−10° 30'	PA 338°; Sep. 4.2"

[4] The subscript indicates the star is variable.

Deep-Sky Objects

Designation	Alternate name	Vis. mag	RA	Dec.	Description
NGC 6694	Messier 26	8.0	18^{h}45.2^m	–09° 27'	Open cluster
NGC 6705	Messier 11/ Wild Duck Cluster	5.8	18^{h}51.1^m	–06° 16'	Open cluster
NGC 6712		8.2	18^{h}53.1^m	–08° 42'	Globular cluster
IC 1295	PK25+4.2	15.0p	18^{h}54.6^m	–08° 50'	Planetary nebula
IC 1287		–	18^{h}30.4^m	–10° 48'	Reflection nebula
Barnard 103		–	18^{h}39.4^m	–06° 41'	Dark nebula
Barnard 110		–	18^{h}50.1^m	–04° 48'	Dark nebula
Barnard 111		–	18^{h}50.1^m	–04° 48'	Dark nebula

Objects in Aquila

Stars

Designation	Alternate name	Vis. mag	RA	Dec.	Description
Alpha (α) Aquilae	Altair	0.76$_v$	19^{h}50.8^m	+08° 52'	12th brightest star
V Aquilae		6.6–8.4	19^{h}04.4^m	–05° 41'	Variable star
R Aquilae		5.5–12.0	19^{h}06.4^m	+08° 14'	Variable star
Eta (η) Aquilae		3.87	19^{h}53.5^m	+01° 00'	Variable star
11 Aquilae		5.2, 8.7	18^{h}59.1^m	+13° 37'	PA 286°, Sep. 17.5"
23 Aquilae		5.3, 9.3	19^{h}18.5^m	+01° 05'	PA 5°, Sep. 3.1"
Σ 2404	Struve 2404	6.9, 8.0	18^{h}50.8^m	+10° 59'	PA 181°, Sep. 3.5"
Zeta (ζ) Aquilae		3.0, 12.0	19^{h}05.4^m	+13° 52'	PA 53°, Sep. 6.5"
Chi (χ) Aquilae		5.6, 6.8	19^{h}42.6^m	+11° 49'	PA 77°, Sep. 0.45"
Pi (π) Aquilae		6.1, 6.9	19^{h}48.7^m	+11° 49'	PA 110°, Sep. 1.4"
Σ 2587	Struve 2587	6.7, 9.4	19^{h}51.4^m	+04° 05'	PA 100°, Sep. 4.4"
β 57		6.3, 10.7	20^{h}05.4^m	+15° 30'	PA 120°, Sep. 2.4"

Deep-Sky Objects

Designation	Alternate name	Vis. mag	RA	Dec.	Description
NGC 6709	Collinder 392	6.7	$18^h 51.5^m$	+10° 21'	Open cluster
NGC 6755		7.5	$19^h 07.8^m$	+04° 14'	Open cluster
NGC 6738		8.3p	$19^h 01.4^m$	+11° 36'	Open cluster
NGC 6773		–	$19^h 15.0^m$	+04° 53'	Open cluster
NGC 6795		–	$19^h 26.0^m$	+03° 31'	Open cluster
NGC 6749		12.4	$19^h 05.1^m$	+01° 47'	Globular cluster
NGC 6760		9.1	$19^h 11.2^m$	+01° 02'	Globular cluster
Pal 11	Palomar 11	12.0	$19^h 45.3^m$	-08° 02'	Globular cluster
Barnard 142		–	$19^h 41.0^m$	+10° 31'	Dark nebula
Barnard 143		–	$19^h 41.4^m$	+11° 01'	Dark nebula
Barnard 133		–	$19^h 06.1^m$	-06° 50'	Dark nebula
Sh (Sharpless) 2–71	PK 036–1.1	13.2	$19^h 02.0^m$	+02° 09'	Planetary nebula
NGC 6751	PK 29–5.1	11.9	$19^h 05.9^m$	-06° 00'	Planetary nebula
NGC 6772	Herschel 14	12.7	$19^h 14.6^m$	-02° 42'	Planetary nebula
NGC 6778	PK 34–6.1	12.3	$19^h 18.4^m$	-01° 36'	Planetary nebula
NGC 6781	Herschel 743	11.4	$19^h 18.4^m$	+06° 33'	Planetary nebula
NGC 6790		10.5	$19^h 23.2^m$	+01° 31'	Planetary nebula
NGC 6803		11.4	$19^h 31.3^m$	+10° 03'	Planetary nebula
SS 433	V1343 Aql	13.0–15.0	$19^h 11.8^m$	+04° 59'	Strange binary star

Objects in Hercules

Stars

Designation	Alternate name	Vis. mag	RA	Dec.	Description
95 Herculis		5.0, 5.1	$18^h 01.5^m$	+21° 36'	PA 258°; Sep. 6.3"

Objects in Sagitta

Stars

Designation	Alternate name	Vis. mag	RA	Dec.	Description
WZ Sagittae		7.0–15.5	$19^h 53.1^m$	+18° 20'	Recurring novae
FG Sagittae		9.5–13.7	$20^h 11.9^m$	+20° 20'	Variable star
V Sagittae		9.5–13.9	$20^h 20.3^m$	+21° 06'	Variable star
U Sagittae		6.6–9.2	$19^h 18.8^m$	+19° 35'	Variable star
Zeta (ζ) Sagittae		5.5, 8.7	$19^h 49.0^m$	+19° 09'	PA 311°; Sep. 8.6"
Theta (θ) Sagittae		6.5, 9.0	$20^h 09.9^m$	+20° 55'	PA 325°; Sep. 11.9"

Deep-Sky Objects

Designation	Alternate name	Vis. mag	RA	Dec.	Description
NGC 6873		6.4	$20^h 08.3^m$	+21° 06'	Open cluster
Harvard 20		7.7	$19^h 53.1^m$	+18° 20'	Open cluster
NGC 6838	Messier 71	8.0	$19^h 53.8^m$	+18° 47'	Globular cluster
IC 4997	PK58–10.1	10.5	$20^h 20.2^m$	+16° 45'	Planetary nebula
NGC 6879	PK58–8.1	12.5	$20^h 10.5^m$	+16° 55'	Planetary nebula
NGC 6886	PK60–7.2	11.4	$20^h 12.7^m$	+19° 59'	Planetary nebula

Objects in Delphinus

Stars

Designation	Alternate name	Vis. mag	RA	Dec.	Description
Beta (β) Delphini		4.0, 4.9	$20^h 37.5^m$	+14° 36'	PA 268°; Sep. 9.6"
Gamma (γ) Delphini		4.3, 5.1	$20^h 45.7^m$	+16° 07'	PA 343°; Sep. 0.5"
U Delphini		7.6, 8.9	$20^h 45.5^m$	+18° 05'	Variable star

Deep-Sky Objects

Designation	Alternate name	Vis. mag	RA	Dec.	Description
NGC 6891	PK54-12.1	10.5	$20^h 15.2^m$	+12° 42'	Planetary nebula
NGC 6905	Blue Flash Nebula	11.1	$20^h 22.4^m$	+20° 05'	Planetary nebula
NGC 6950		—	$20^h 41.2^m$	+16° 38'	Open cluster
Harrington 9		—	$20^h 38^m$	+13° 30'	Open cluster

Objects in Vulpecula

Stars

Designation	Alternate name	Vis. mag	RA	Dec.	Description
β 441		6.2, 10.7	$20^h 17.5^m$	+29° 09'	PA 66°; Sep. 5.9"
Σ 2445	Struve 2445	7.2, 8.9	$19^h 04.6^m$	+23° 20'	PA 263°; Sep. 12.6"

Deep-Sky Objects

Designation	Alternate name	Vis. mag	RA	Dec.	Description
Collinder 399	Coathanger or Brocchi's Cluster	3.6	$19^h 25.4^m$	+20° 11'	Open cluster
Stock 1		5.3	$19^h 35.8^m$	+25° 13'	Open cluster
NGC 6802	Herschel 14	8.8	$19^h 30.6^m$	+20° 16'	Open cluster
NGC 6823	Herschel 18	7.1	$19^h 43.1^m$	+23° 18'	Open cluster
NGC 6820			$19^h 43.1^m$	+23° 17'	Emission nebula
NGC 6882	Herschel 22	8.1	$20^h 11.7^m$	+26° 33'	Open cluster
NGC 6885	Caldwell 37	5.7	$20^h 12.0^m$	+26° 29'	Open cluster
NGC 6940	Herschel 8	6.3	$20^h 34.6^m$	+28° 18'	Open cluster
NGC 6842	PK65+0.1	13.1	$19^h 55.0^m$	+29° 17'	Planetary nebula
NGC 6853	Messier 27/Dumbbell Nebula	7.3	$19^h 59.6^m$	+22° 43'	Planetary nebula

Objects in Cygnus

Stars

Designation	Alternate name	Vis. mag	RA	Dec.	Description
Beta (β) Cygni	Albireo	3.1, 5.1	19^{h}30.7^m	+27° 58'	PA 54°; Sep. 34"
Delta (δ) Cygni		2.9, 6.3	19^{h}45.0^m	+45° 08'	PA 221°; Sep. 2.5"
17 Cygni		5.0, 8.5	19^{h}46.4^m	+33° 44'	PA 69°; Sep. 26"
49 Cygni		5.6, 7.9	20^{h}41.0^m	+32° 18'	PA 45°; Sep. 2.5"
OΣ 437		6.2, 6.9	21^{h}20.8^m	+32° 27'	PA 24°; Sep. 2.5"
B 677		4.9, 9.9	20^{h}47.2^m	+34° 22'	PA 121°; Sep. 9.9"
OΣ 390		6.5, 9.3, 11.1	19^{h}55.1^m	+30° 12'	PA 22°; Sep. 9.5"^{AB} / PA 175°; Sep. 6.3"^{AC}
RS Cygni		6.5–9.3	20^{h}13.4^m	+38° 44'	Variable star
Chi (χ) Cygni		3.3, 14.2	19^{h}50.6^m	+32° 55'	Variable star
Alpha (α) Cygni	Deneb	1.25	20^{h}41.3^m	+45° 15'	20th brightest star
61 Cygni		5.20, 6.05	21^{h}06.9^m	+38° 45'	PA 195°; Sep. 321"
16 Cygni B		6.25	19^{h}41.8^m	+50° 31'	Possible planetary system

Deep-Sky Objects

Designation	Alternate name	Vis. mag	RA	Dec.	Description
NGC 6819	Collinder 403	7.3	19^{h}41.3^m	+40° 11'	Open cluster
NGC 6871	Collinder 413	5.2	20^{h}05.9^m	+35° 47'	Open cluster
Roslund 5			20^{h}10.0^m	+33° 46'	Open cluster
NGC 6913	Messier 29	6.6	20^{h}23.9^m	+38° 32'	Open cluster
NGC 7092	Messier 39	4.6	21^{h}32.2^m	+48° 26'	Open cluster
Ruprecht 173	–		20^{h}41.8^m	+35° 33'	Open cluster
NGC 6894	PK69–2.1	12.3	20^{h}16.4^m	+30° 34'	Planetary nebula
NGC 7026	PK89+0.1	10.9	21^{h}06.3^m	+47° 51'	Planetary nebula
NGC 7027	PK84+3.1	8.4	21^{h}07.1^m	+42° 14'	Planetary nebula
NGC 7048		12.1	21^{h}14.2^m	+46° 16'	Planetary nebula
NGC 6826	Caldwell 15/Blinking Planetary	8.8	19^{h}44.8^m	+50° 31'	Planetary nebula

Objects in Cygnus (continued)

Stars

Designation	Alternate name	Vis. mag	RA	Dec.	Description
Campbell's Hydrogen Star		11.3	19^{h}34.8^m	+30° 31'	PK 64 + 5.1 Planetary nebula
NGC 7000	Caldwell 20/North America Nebula	–	20^{h}58.8^m	+44° 20'	Emission nebula
IC 5067/70	Pelican Nebula	–	20^{h}50.8^m	+44° 21'	Emission nebula
NGC 6888	Caldwell 27/Crescent Nebula	–	20^{h}12.0^m	+38° 21'	Emission nebula
IC 5146	Caldwell 19/Cocoon Nebula	–	21^{h}53.4^m	+47° 16'	Emission nebula
Barnard 168		–	21^{h}53.3^m	+47° 12'	Dark nebula
CRL 2688	Egg Nebula	10.7	21^{h}00.6^m	+54° 33'	Protoplanetary nebula
Barnard 145		–	20^{h}02.8^m	+37° 40'	Dark nebula
Harrington 10		–	21^{h}00^m	+55°	Dark nebula
NGC 6960	Caldwell 34/Veil Nebula (Western section)		20^{h}45.7^m	+30° 43'	SNR & emission nebula
NGC 6992	Caldwell 33/Veil Nebula (Eastern section)		20^{h}56.4^m	+31° 43'	SNR & emission nebula
NGC 6974–79	Veil Nebula (Central section)		20^{h}50.8^m	+31° 52'	SNR & emission nebula

Objects in Lyra

Stars

Designation	Alternate name	Vis. mag	RA	Dec.	Description
Beta (β) Lyrae	Shellak	3.4–4.4	$18^h 50.1^m$	+33° 22'	Variable star
Σ 2470	Struve 2470	6.6, 8.6	$19^h 08.8^m$	+34° 46'	PA 271°; Sep. 13.4"
Σ 2474	Struve 2474	6.7, 8.8	$19^h 09.1^m$	+34° 36'	PA 262°; Sep. 16.2"
Eta (η) Lyrae		4.4, 9.1	$19^h 13.8^m$	+39° 09'	PA 81°; Sep. 28"
RR Lyrae		7.0–8.1	$19^h 25.5^m$	+42° 47'	Variable star

Deep-Sky Objects

Designation	Alternate name	Vis. mag	RA	Dec.	Description
Stephenson 1		3.8	$18^h 53.5^m$	+36° 55'	Open cluster
NGC 6791		9.7	$19^h 20.7^m$	+37° 51'	Open cluster
NGC 6779	Messier 56	8.3	$19^h 16.6^m$	+30° 11'	Globular cluster
NGC 6720	Messier 57/Ring Nebula	8.8	$18^h 53.6^m$	+33° 02'	Planetary nebula

Objects in Lacerta

Stars

Designation	Alternate name	Vis. mag	RA	Dec.	Description
Σ 2876	Struve 2876	7.8, 9.3	22^{h}12.0^m	+37° 39'	PA 68°; Sep. 11.8"
Σ 2894	Struve 2894	6.1, 8.3	22^{h}18.9^m	+37° 46'	PA 194°; Sep. 15.6"
Σ 2902	Struve 2902	7.6, 8.5	22^{h}23.6^m	+45° 21'	PA 89°; Sep. 6.4"
8 Lacertae	Σ 2922	5.7, 6.5	22^{h}35.8^m	+39° 38'	PA 186°; Sep. 22.4"

Deep-Sky Objects

Designation	Alternate name	Vis. mag	RA	Dec.	Description
NGC 7209	Herschel 53	7.7	22^{h}05.2^m	+46° 30'	Open cluster
NGC 7243	Caldwell 16	12.2	22^{h}10.2^m	+41° 00'	Open cluster
IC 1434	Collinder 445	9.0	22^{h}10.5^m	+52° 50'	Open cluster
IC 5217		11.3	22^{h}23.9^m	+50° 58'	Open cluster

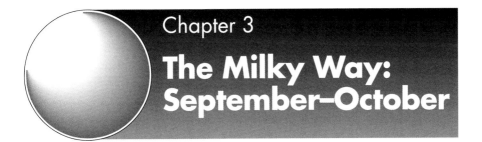

Chapter 3

The Milky Way: September–October

Cepheus, Andromeda, Camelopardalis, Cassiopeia
R.A 2^h to 5^h; Dec. 50° to75°; Galactic longitude[1] 105° to 140° ; Star Chart 3

3.1 Cepheus

We now begin to look at those constellations which ride high in the sky for northern observers, but may be rather low, or even unobservable, for southern observers. In fact, several of the constellations are circumpolar for northern observers, which in theory means you could observe them on any night of the year, although there will be times when they are very low in the sky, and so atmospheric extinction will hinder your view. Let's begin looking at our collection of autumn Milky Way constellations (see Star Chart 3).

Cepheus is one of those constellations that do not immediately jump out at you, and indeed its brightest star is at magnitude 2.5. It lies at the edge of the Milky Way and older Star Charts will show only its southern and westernmost areas which are in fact located in the Milky Way. However more up-to-date atlases show that most of its southern reaches are in the Milky Way.

At a casual glance **Cepheus** looks quite empty but don't let this fool you, as it is actually full of delights (see Star Chart 3.1). A few are visible to the naked eye, and there are ample objects for small telescopes. In addition it also has a lot of objects that are more suited for large aperture telescope of, say, 30 cm and more, but we will not concern ourselves too much with those faint objects. Alas, for most southern observers this is just a constellation to read about. The center of Cepheus transits at the end of September, even though most of what we will look at can be seen at an earlier date. Nevertheless, it is placed in this chapter for accuracy and completeness.

There are some nice stars in Cepheus, so let's start with a superb example of a colored star. This is **Mu (μ) Cephei**, located on the northeastern edge of the nebulosity IC1396. It is also known as the **Garnet Star**, so-named by William Herschel, and is one of the reddest stars in the entire sky. It has a deep orange or red color seen against a backdrop of faint white stars. In medium-aperture telescopes it looks more of an orange-red color, and in large telescopes it changes to yellowish orange whereas in small telescopes it does indeed

---

[1] See Appendix 1 for details on astronomical coordinate systems.

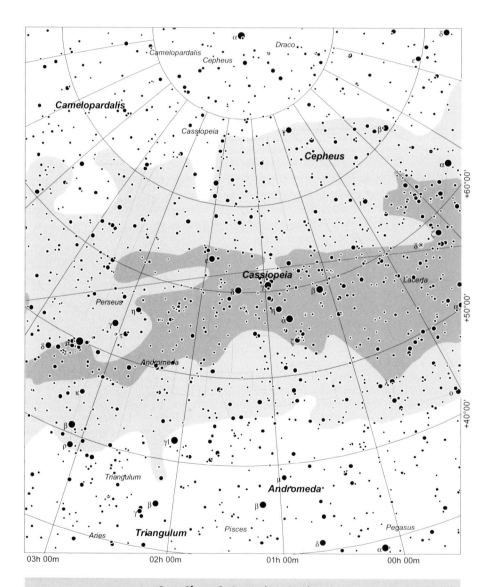

Star Chart 3. September–October.

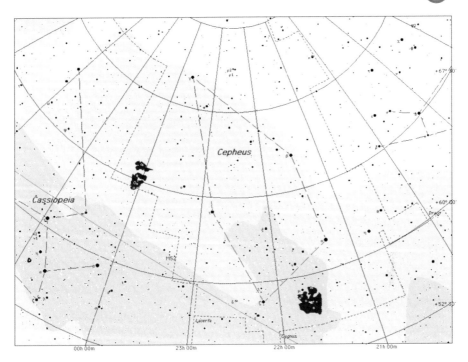

Star Chart 3.1. Cepheus.

have a lovely deep red color, especially when it is near its minimum magnitude. It is a pulsating red giant star, with a period of about 730 days, varying from 3.4 to 5.1 magnitude.

A nice double star to begin with is **Beta (β) Cephei**, shining at magnitudes 3.2 and 7.9. The system consists of nice white and blue stars. Using a large-aperture telescope, the secondary takes on a definite green tint. It is also a Cepheid variable with very small light variations. Another fine double is **OΣ 440**, which is a pair of 6th and 10th magnitude stars that are a lovely orange and blue color. A fine system for small telescopes that is also a true binary star system is **Xi (ξ) Cephei**. In small telescopes of, say, 8 cm, the stars both appear white, although some observers see one of the stars as a pale reddish color, whereas in larger telescopes the stars take on definite tints of yellow and red. What makes the star interesting for me is that it may be an outlying member of the **Taurus Stream**. This is a truly vast ensemble of stars that originally had a common origin, but over time have evolved and spread throughout space, so that now the only common factor between them is their motion through space. The stream extends to over 200 light years beyond the Hyades star cluster, and 300 light years behind the Sun. Thus, the Sun is believed to lie within this stream. Capella and Alpha α Canum Venaticorum are also thought to be members of the large Taurus Stream, which has a motion through space similar to the Hyades star cluster, and thus may be related.

Two more doubles warrant our attention: Krueger 60 and Delta (δ) Cephei. **Krueger 60** is a faint double star that is also one of the nearest binary systems to us, at a distance of just over 13 light years. The two stars, of magnitudes 9.8 and 11.4 and separated by about 3.5 arcseconds, are red dwarf stars and their masses are a fraction of that of the Sun. In fact **Kr60^B**, as it is called, has one of the smallest masses known. A telescope of at least 15 cm aperture is needed in order to resolve the system, and a high magnification will also help, when they will appear as two very distinctively red stars.

Delta (δ) Cephei, on the other hand, is the prototype star of the classic short-period pulsating variables known as Cepheids. It was first discovered in 1784 by the British amateur John Goodricke. It is an easy favorite with amateurs as four bright stars also lie in the vicinity – **Epsilon (ε) Persei** (4.2 mag), **Zeta¹ (ζ) Persei** (3.4 mag), **Zeta (ζ) Cephei** (3.35 mag), and **Eta (η) Cephei** (3.43 mag). The behavior of the star is as follows: it will brighten for about $1\frac{1}{2}$ days, and will then fade for 4 days, with a period of 5 days, 8 hours and 48.2 minutes. Its magnitude range is from 3.48 to 4.37. Delta Cephei is also a famous double star, with the secondary star (6.3 mag), a nice white color, which contrasts well with the yellowish tinge of the primary.

Following on from the original Cepheid variable star is another type of variable, an eclipsing binary star, **U Cephei**. The brighter of the pair is occulted by the fainter companion every 2.5 days. This results in a four-hour fall in magnitude, from 6.7 to 9.2, which is then followed by a two-hour eclipse. It is a nice object for small telescopes and perhaps large binoculars. What makes the star so interesting is that, because of their closeness, the bright star is actually ripping material from the fainter star. This results in the fainter star losing a substantial amount of its mass, causing the orbital period to slow down.

As I mentioned above, to the naked eye there doesn't seem a lot in Cepheus, but it has a lot of clusters, a few nebulae and even a respectable number of galaxies. Due to the effect of interstellar dust, however, some of the clusters and most of the galaxies are faint and small. So we shall concern ourselves just with those objects that are big and bright!

Our first extended object is the open cluster **NGC 6939**. This is a moderately bright and small cluster at magnitude 7.8, but is unresolvable in binoculars (see Star Chart 3.2) – a challenge, as the brightest member is only of 11.9 magnitude. In telescopes of aperture 10 cm, it will appear as a small hazy spot with just a few very faint stars resolved (see

Star Chart 3.2. NGC 6939.

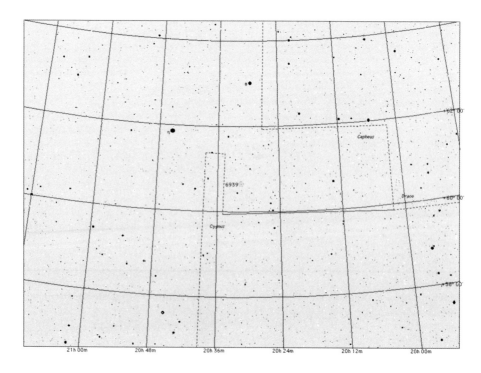

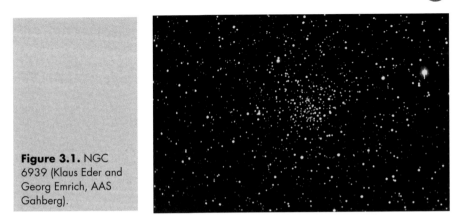

Figure 3.1. NGC 6939 (Klaus Eder and Georg Emrich, AAS Gahberg).

Figure 3.1). With larger apertures many stars can be resolved into arcs and chains that fade into the background of the Milky Way. What is special, however, is that a large number of stars are packed into a tiny area, some 8 arcminutes, making this one of the richest open clusters in the northern sky.

Located in the same field of view is the galaxy **NGC 6946,** which is a face-on spiral, but a challenge to locate in small apertures.

A cluster that is a highlight of Cepheus is **IC 1396**. Although a telescope of at least 20 cm is needed to really appreciate this cluster, it is nevertheless worth searching out (see Star Chart 3.3). It can be seen in telescopes as small as 8 cm, when it will appear as a circular

Star Chart 3.3. IC 1396.

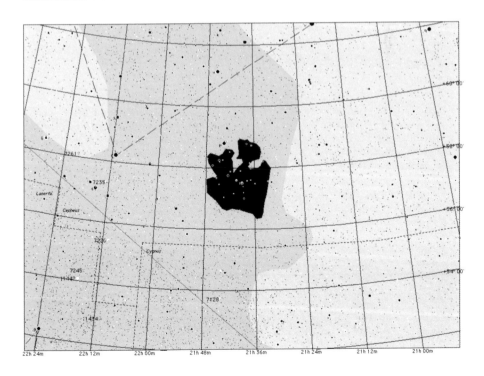

Figure 3.2. IC 1396 (Matt BenDaniel, http://starmatt.com).

hazy patch. It can even be seen under superb conditions with the naked eye (see Figure 3.2). It lies south of Herschel's Garnet Star and is rich but compressed. What makes this so special, however, is that it is cocooned within a very large and bright nebula. I say bright with some caution as it is notoriously difficult to see, as it depends on observing conditions, optics, etc. but an [OIII] filter will of course help in its detection. Because of its large size, the nebula is best seen with binoculars.

A couple of nice clusters suitable for telescopes of aperture 20 cm are **NGC 7142** and **NGC 7160** (see Star Chart 3.4). The former is a rich cluster of about 40 stars covering an area of 10 arcminutes. The stars are mostly of 12–14th magnitude, but this makes for an integrated magnitude of about 9.4 (see Figure 3.3). The latter cluster is brighter at magnitude 6.1 but has fewer stars in small-aperture telescopes and covers a smaller area of the sky, some 7 arcminutes (see Figure 3.4). Both of these clusters take a high magnification well.

Another fine pair of clusters are **NGC 7235** and **NGC 7261**, both situated amongst lovely star fields of the Milky Way (see Star Chart 3.5). NGC 7235 is a nice group of about 20 stars of 9–12th magnitude. If a larger telescope is used, many more cluster members become resolved (see Figure 3.5). NGC 7261 lies 1° east of **Zeta (ζ) Cephei** and consists of about 20

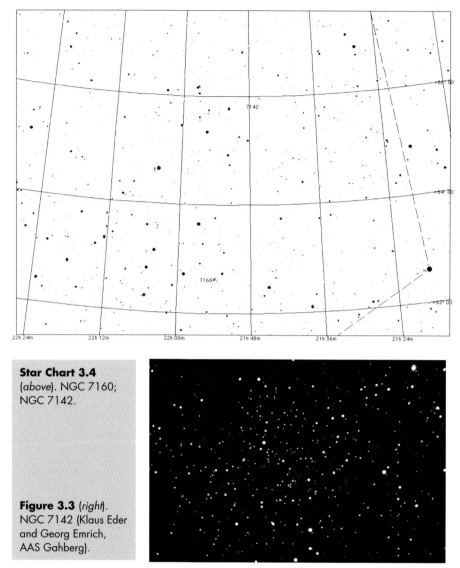

Star Chart 3.4 (*above*). NGC 7160; NGC 7142.

Figure 3.3 (*right*). NGC 7142 (Klaus Eder and Georg Emrich, AAS Gahberg).

stars loosely scattered over an area of about 6 arcseconds. There are several delightful star chains visible here as well (see Figure 3.6).

One cluster that should perhaps be reclassified as an asterism is **NGC 7281**. This is a fairly loose group of about 25 stars in a triangular shape (see Figure 3.7). In fact, it doesn't show up as such on many star atlases. Do you see it as a cluster?

Appearing as just a hazy object in telescopes of 20 cm aperture is the fairly unknown open cluster **King 10**. With a low magnification only about ten stars will be seen set against the unresolved cluster members. But higher magnification and aperture will reveal dozens more in a lovely star field. This is a cluster that is a nice surprise and worth seeking out. A really beautiful cluster is **NGC 7510** (see Star Chart 3.7). It is rich and bright at 8th magnitude, and is highly compact, as it is only in an area some 4 arcminutes across (see Figure

Figure 3.4. NGC 7160 (Space Telescope Science Institute, AAO, UK–PPARC, ROE, National Geographic Society, and California Institute of Technology).

Star Chart 3.5. NGC 7235; NGC 7261.

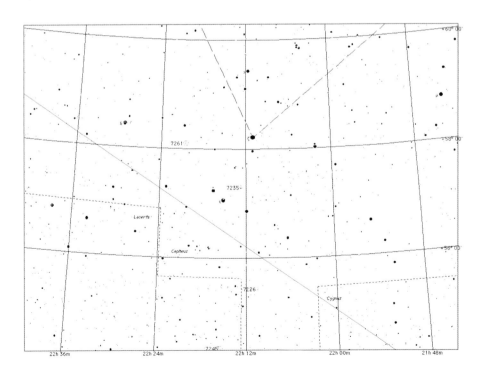

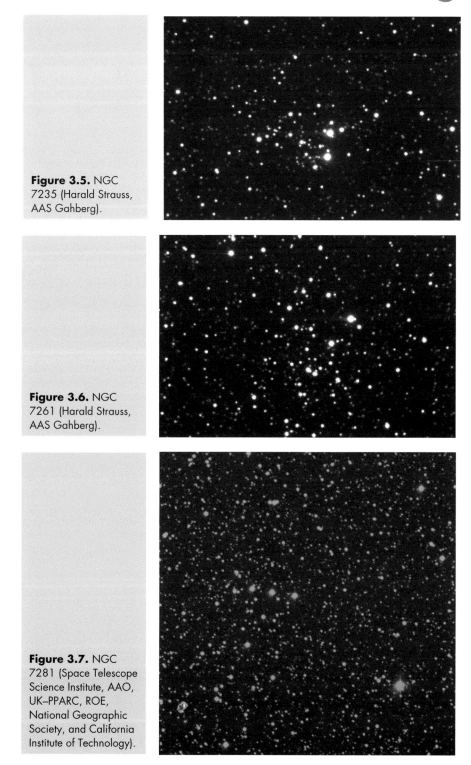

Figure 3.5. NGC 7235 (Harald Strauss, AAS Gahberg).

Figure 3.6. NGC 7261 (Harald Strauss, AAS Gahberg).

Figure 3.7. NGC 7281 (Space Telescope Science Institute, AAO, UK–PPARC, ROE, National Geographic Society, and California Institute of Technology).

Figure 3.8. NGC 7510 (Harald Strauss, AAS Gahberg).

3.8). With a high magnification a lot more can be seen, and large-telescope owners should make an effort to seek this out as it really is a lovely object, set as it is in the Milky Way.

Near to NGC 7510 is our last open cluster, **Markarian 50.** In small telescopes of about 10 cm it will appear as a small group of about six stars that seem to be set within a faint, almost unresolved or imagined nebulosity. Larger telescopes will reveal several more stars that are once again just at the limit of resolution.

Let's now turn our attention to some nebulae. Although quite inconspicuous, dark nebulae have a presence here, and scanning with telescopes and perhaps large binoculars will reveal these hitherto unnoticed objects. Most prominent of these are **Barnard 169, Barnard 170** and **Barnard 171**; collectively they are also known as **Lynds 1151**. Barnard 171 is a dense area of dark cloud that contrasts rather well with the surrounding Milky Way. Irregularly shaped, it has several extensions that branch off. The northwest extension contains the remaining two Barnard object, B 169 – 170. Then there are **Barnard 173** and **Barnard 174**, collectively known as **Lynds 1164**. The whole object is a nice dark cloud that has a very distinctive S-shape to it; the southern part is B 173 and contrasts well with the Milky Way, but B 174 has a lovely scattering of stars over it, which diminishes its appeal. Nevertheless it is in a wonderfully rich star field. Remember that observing dark nebulae requires the darkest and most transparent of nights.

Now for something that can be seen with the naked eye. **Harrington 11** is the bright and very conspicuous band of starlight that seems to have broken away from the Milky Way and has veered north towards to the southwest of the constellation. It will appear as a straight patch of light about $10° \times 5°$. It is part of the **Cepheus OB2** association, and when observed in any optical equipment will fragment into many splendid star fields. Often ignored, it is well worth the time to seek it out.

There are a lot of both emission and reflection nebulae in Cepheus, but, alas, most of them are beyond the reach of small telescopes, and by that I mean anything less than 30 cm aperture. But there are a few planetary nebulae so let's look at them. The planetary nebula **NGC 40 (Caldwell 2)** is a spectacular object, often overlooked, probably because it is at 12th magnitude (see Star Chart 3.6). Appearing as a star in binoculars, it needs an aperture of at least 20 cm for its planetary nebula nature to become apparent (see Figure 3.9). It is bright[2] and oval-shaped, and has brighter regions still at its west and east sections, and has a lighter northern area, but this latter feature is seen only under perfect seeing conditions. In large telescopes it has a definite blue-green color and I think it is an object that is worth seeking out, as it is very spectacular, but sadly is frequently passed over and not mentioned very often in other star guides.

---

[2] Bright when it is seen in a telescope of aperture 20 cm and greater!

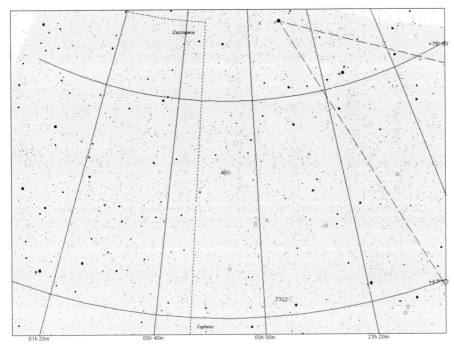

Star Chart 3.6 (*above*). NGC 40.
Star Chart 3.7 (*below*). NGC 7510; NGC 7354.

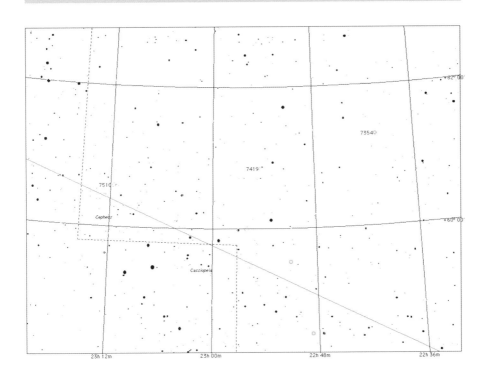

Figure 3.9. NGC 40 (Klaus Eder and Georg Emrich, AAS Gahberg).

Another nice planetary nebula is **NGC 7354**, which lies in a triangle of three 10th magnitude stars (see Star Chart 3.7). However, it is small at only 20 arcseconds and faint at magnitude 12.2, so finding it can be a problem. Even a telescope of 25 cm will not show a central star, and its appearance is not guaranteed with even larger apertures (see Figure 3.10).

Our final planetary, **NGC 7139** (**PK 104.7**), and indeed final object, is a difficult, although not impossible, nebula to locate and observe (see Star Chart 3.8), the reason being its magnitude of 13.3, even though it is large, with a diameter of nearly 80 arcseconds

Figure 3.10. NGC 7354 (Klaus Eder and Georg Emrich, AAS Gahberg).

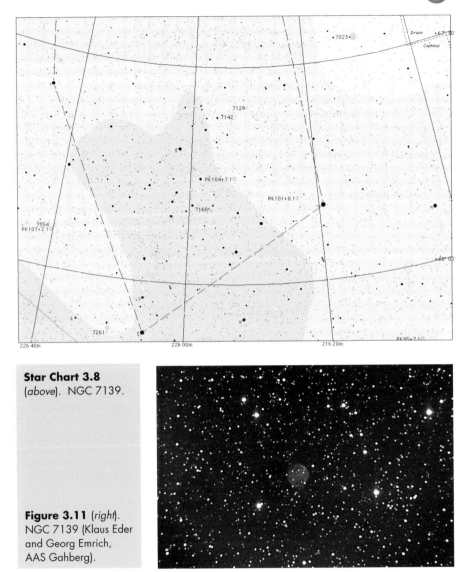

Star Chart 3.8
(*above*). NGC 7139.

Figure 3.11 (*right*).
NGC 7139 (Klaus Eder
and Georg Emrich,
AAS Gahberg).

(see Figure 3.11). It will appear as a faint greyish patch, and has an extremely faint central star, magnitude 18.8.

Even though there are not too many objects in this part of the Milky Way, Cepheus is still a great constellation for scanning the Milky Way in the late autumn evenings.

3.2 Andromeda

It came as a surprise to me to discover that this constellation has the Milky Way passing through its northern regions, as I had always assumed it was far from it. This means that

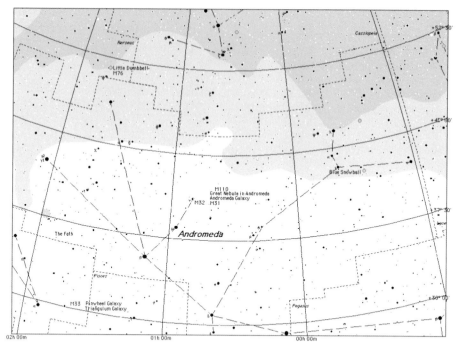

Star Chart 3.9. Andromeda.

Andromeda has several fine objects that we can observe (see Star Chart 3.9). Alas, however, the boundary of the Milky Way does not encompass Andromeda's most famous resident, Messier 31, the Andromeda Galaxy, so this is the only mention it will receive. Some parts of the constellation are visible to southern observers, but they will be low on the horizon.

There are not many stars that we can observe that are of interest to us in Andromeda, but there are a couple, including one of the loveliest double stars in the entire sky. One of these stars is **Groombridge 34**. This is a red dwarf binary system that is one of the closest to us at nearly 12 light years. It is a relatively easy system to resolve and will show a bright pair of red stars shining at magnitudes 8.2 and 10.6. What makes it interesting to amateur astronomers, however, is that it has a large proper motion that can be plotted over several years. If you carefully draw the two stars, and then observe some time later you will see that their positions have changed against the background star field.

Our next star is the magnificent **Gamma (γ) Andromedae**. This is a superbly colored system comprising a golden primary and a greenish-blue secondary. One of the best doubles in the sky, it benefits from having your optical system slightly out of focus, as this will enhance the color contrast. The star is easily resolved in small telescopes, but the companion star is itself a double that will need larger apertures in order to be split. The primary is itself a spectroscopic binary, so what we have here is a quadruple star system. Wonderful!

The one open cluster we can observe is **NGC 7686 (Herschel 69)** (see Star Chart 3.10). This is a sparse and widely dispersed cluster containing many 10th and 11th magnitude stars. It can be seen in telescopes of aperture 20 cm but is best seen with large-aperture telescopes (see Figure 3.12).

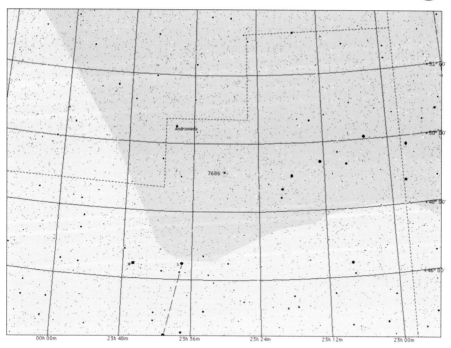

Star Chart 3.10. NGC 7686.

Figure 3.12. NGC 7686 (Space Telescope Science Institute, AAO, UK–PPARC, ROE, National Geographic Society, and California Institute of Technology).

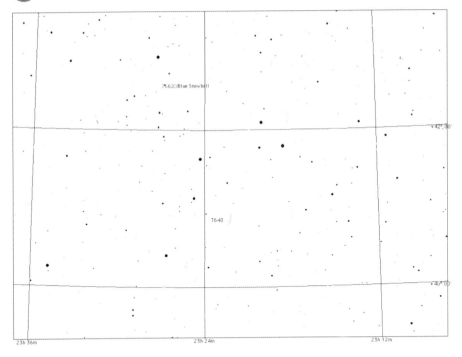

Star Chart 3.11. NGC 7662; NGC 7640.

There is one object that we can observe that is a highlight of Andromeda, the planetary nebula **NGC 7662 (Caldwell 22)**, or, as it is sometimes known, the **Blue Snowball** (see Star Chart 3.11 and Figure 3.13). This is a nice planetary nebula that is even visible in binoculars owing to its striking blue color, but even then it will only appear stellar-like. In tele-

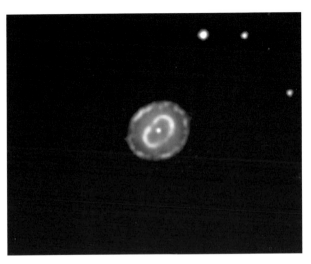

Figure 3.13. NGC 7662 (Robert Schulz, AAS Gahberg).

Figure 3.14. NGC 7640 (Harald Strauss, AAS Gahberg).

scopes of 20 cm, the disk is seen, along with some ring structure. With a larger aperture, subtle color variations appear – blue-green shading. The central star will need large apertures in order to be seen, and even then it will only be glimpsed from time to time under the best seeing conditions. Research indicates that the planetary nebula has a structure similar to that seen in the striking *Hubble Space Telescope* (HST) image of the Helix Nebula, showing Fast Low-Ionization Emission Regions (fliers). These are clumps of above-average-density gas ejected from the central star before it formed the planetary nebula.

Our final object on our brief visit to Andromeda is the galaxy **NGC 7640** (see Star Chart 3.11). This is a rather faint, 11th magnitude galaxy that in small telescopes will appear as an elongated smudge some 7 × 2 arcseconds that may exhibit a slight brightening at its center (see Figure 3.14). Larger apertures will just magnify and brighten the same image but some of the halo will become apparent.

3.3 Camelopardalis

As one of the largest constellations in square degrees, and being circumpolar for northern observers, you would think that this would be a well-recognized and much-observed constellation, whereas in fact **Camelopardalis** is probably one of the least-observed and least-known constellations in the entire sky (see Star Chart 3.12). The reason for this is simple: there are no stars brighter than 4th magnitude in it, and its shape is meandering and nondescript. Its nonobservability actually belies the fact that it has quite a lot of objects well worth observing, particularly galaxies and clusters.

Its center transits in late December, but the Milky Way and the galactic equator just pass through its southeastern regions, and so I have decided to place it here, instead of a later chapter. Furthermore, although it should come after Cassiopeia in this chapter, I have decided to keep that section till last, as it is full of celestial treasures. For southern observers, it is nearly, or completely, unobservable, depending where you are in the southern hemisphere.

There are, as usual, some nice double stars to observe. Of these, Σ (**Struve**) **390**, Σ (**Struve**) **485** and Σ (**Struve**) **550** (**1 Camelopardalis**), are within the boundaries of the Milky Way. The first is a nice system of unequally bright stars, one that is white, and one that takes on a purple tint, a color that is not often seen in stellar systems. It can be easily seen in telescopes of about 10 cm and greater. The fainter member is also an eclipsing

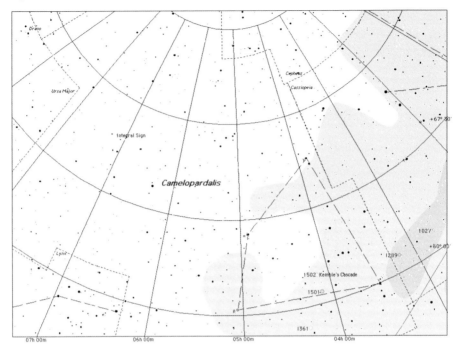

Star Chart 3.12. Camelopardalis.

binary of the Algol type. However, the magnitude change is only about one-quarter of a magnitude so it is slightly difficult to observe with the naked eye, although I imagine not impossible. The second system, Σ (Struve) 485, will need a slightly bigger aperture, say 20 cm or more, in order to be reasonably seen, and will appear as two nicely colored stars, both blue-white. The final star, Σ (Struve) 550, is once again an easy double of a white star and a pale blue one. All the other double stars in Camelopardalis, of which there are quite a few, and within reach of most amateurs, alas all lie outside of the Milky Way.

Our first open cluster is **Stock 23**, which lies close to the Camelopardalis–Cassiopeia border (see Star Chart 3.13). This is a little-known cluster and binoculars will reveal several stars. However, it is best viewed in medium-aperture telescopes where some 40 stars can be seen. It is bright and large but spread out. It lies some 10° to the northwest of **Alpha (α) Persei**. There are a few reports that the cluster has some associated nebulosity. But no one I have spoken to has ever seen it. Have you?

Another open cluster that lies within the Milky Way is **Tombaugh 5**. However, be warned now that it will require a large-aperture telescope in order to be seen in any detail. It is faint, with an integrated magnitude of 8.4, but most of its 40 or so members are 12th and 13th magnitude, hence the warning above. Spread out over a respectable area, it nevertheless stands out well against the star-filled background.

Another nice cluster that is also fine for binoculars is **NGC 1502 (Herschel 47)**. It is bright but oddly enough a problem to locate (see Star Chart 3.14). It can be seen with the naked eye under good conditions, and with binoculars will appear as a hazy round patch of light. It is a rich and bright cluster, but small, and may resemble a fan shape, although this does depend on what the observer sees. In larger telescopes it is a lovely sight, bright, rich and standing out well against the background (see Figure 3.15).

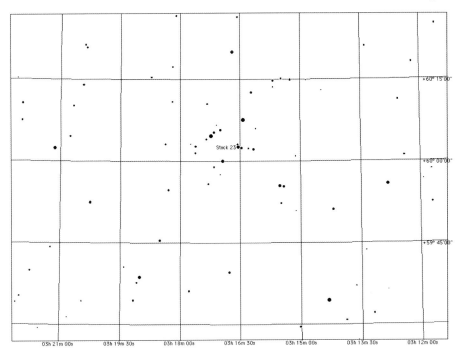

Star Chart 3.13 (*above*). Stock 23.
Star Chart 3.14 (*below*). NGC 1502; Stock 23; NGC 1501.

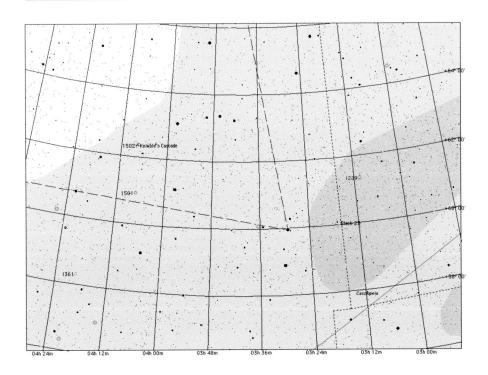

Figure 3.15. NGC 1502 (Harald Strauss, AAS Gahberg).

Also contained in the cluster are two multiple stars Σ **Struve 484** and **485**. The former is a nice triple system, but the latter is a true spectacle with nine components! Seven of these are visible in a telescope of 10 cm aperture, ranging between 7th and 13th magnitude. The remaining two components, of 13.6 and 14.1 magnitude, should be visible in a 20 cm telescope. In addition, the system's brightest compo nent, **SZ Camelopardalis**, is an eclipsing variable star, which changes magnitude by 0.3 over 2.7 days. However, what makes this cluster even nicer is the long string of stars that appears to end at the cluster. The stars in the string, nearly 2° in length, are around 5th to 8th magnitude and lie to the northwest of NGC 1502.

This asterism is called **Kemble's Cascade**, named after the late Lucien Kemble, a Canadian astronomer. The cascade is a grand sight in binoculars. Furthermore, at the end of the cascade and near the cluster is the variable star **UV Camelopardalis**. This is a semiregular variable that varies in brightness from 7.5 to 8.1 magnitude over a period of 294 days.

A planetary nebula is also available for observation: **NGC 1501 (Herschel 53)** (see Star Chart 3.14). It has also been called the **Oyster Nebula** and is a very nice blue planetary nebula that can easily be seen in telescopes of 20 cm, and even glimpsed in apertures of 10 cm (see Figure 3.16). However, with a larger aperture, some structure can be glimpsed, and many observers liken this planetary nebula to that of the Eskimo Nebula. The central star can be seen if a high magnification is used – 300×.

Figure 3.16. NGC 1501 (Harald Strauss, AAS Gahberg).

Several dark nebulae are also within this section of the Milky Way, including one vast area of dust cloud that seems as if part of the Milky Way is missing. This large expanse of dust actually lies within the Perseus–Camelopardalis border.

Although this book is aimed at the small to medium aperture telescope owner, there are within the Milky Way area of Camelopardalis two reflection nebulae that are exceedingly faint and so need large telescopes in order to be seen. These are **van den Bergh 14** and **van den Bergh 15**. To say they are faint is no exaggeration, as they are very difficult objects to see visually. You will need to use averted vision and the sky has to be very dark and very transparent. In both cases they will appear as faint pale streaks of light. Perhaps this is not so much a visual observing challenge, but a photographic or CCD-imaging challenge.

There are many galaxies in Camelopardalis, but none that reside in the area we are concerned about.

3.4 Cassiopeia

I have kept till last the constellation that for many holds the greatest celestial wonders of this season: Cassiopeia. Its very distinctive "W" shape on the sky is almost universally recognized and is truly a treasure chest of clusters, double and multiple stars and nebulae. In fact, it seems as if it has more than its fair share! The Milky Way is especially rich here, and many fruitful hours can be spent just scanning the region with binoculars or a rich-field telescope (see Star Chart 3.15). Such is the plethora of stars amidst the Milky Way's

Star Chart 3.15. Cassiopeia.

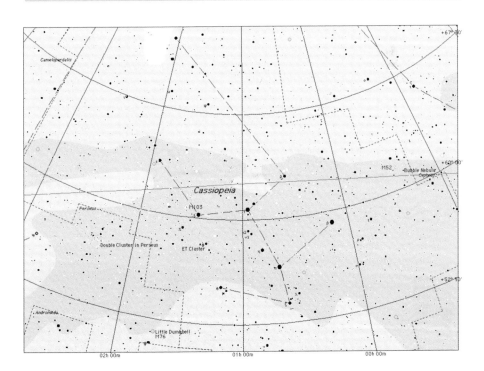

Figure 3.17. Cassiopeia (Matt BenDaniel, http://starmatt.com).

ever-present glow that it is sometimes difficult to discern what constitutes a cluster and what doesn't. The constellation is circumpolar for northern observers, but is probably forever beyond the reach for observers in the southern hemisphere. It transits in early October, but providing you are willing to stay up all night, it can, like so many other Milky Way constellations, be seen during the summer months (see Figure 3.17).

Without further ado, let's start by looking at those objects for which Cassiopeia is rightfully famous – open clusters!

One of the densest clusters that lies north of the celestial equator is **Messier 52 (NGC 7654)** (see Star Chart 3.16). This is a small, rich, and fairly bright cluster with several stars that are visible in binoculars, but telescopic apertures are needed to fully appreciate this cluster. It is one of the few clusters that show a distinct color where any observers report a faint blue tint to the group, and this, along with a fine topaz-colored (blue) star and several nice yellow and blue stars, makes it a very nice object to observe (see Figure 3.18). In medium-aperture telescopes about 80 stars will be seen, while in larger telescopes over 150 cluster members will be on view. Apparently, the cluster has a star density of the order of 50 stars per cubic parsec! Incidentally, the very faint cluster **Czernik 43** can be seen to the south of M52 but only in large telescopes.

For those of you who like a challenge there is **King 12**, a very faint open cluster containing many 10th, 11th and 12th magnitude stars. Close to it, some 10 arcminutes to the southeast, lies **Harvard 21**, an equally faint but poor cluster of five–seven 10th magnitude

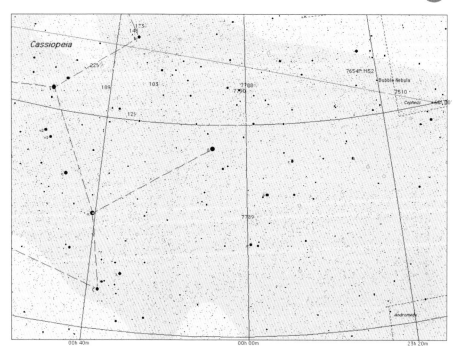

Star Chart 3.16. NGC 103; NGC 129; NGC 139; NGC 229; NGC 7654; NGC 7788; NGC 7789; NGC 7790; Messier 52.

stars.[3] In a similar vein is **NGC 133 (Collinder 3)**, also visible in small telescopes as a handful of 9th magnitude stars. A wonderful cluster that was for some unknown reason omitted from Messier's list is **NGC 7789 (Herschel 30)** (see Star Chart 3.16). It is visible as a hazy spot to the naked eye, and even with small binoculars is never fully resolvable. But through a telescope it is seen as a very rich and compressed cluster and is probably one of the best clusters to observe for telescopes in the 10–15 cm range (see Figure 3.19). With large aperture, the cluster is superb and has been likened to a field of scattered diamond dust. It contains hundreds stars of 10th magnitude and fainter spread over half a degree of sky. Current research indicates that it is one of the oldest open clusters known, some 2 billion years, but of course this is relatively young compared to the globular clusters in the Galaxy.

Another faint cluster that is, however, located in a lovely rich star field is **NGC 7790 (Herschel 56)** (see Star Chart 3.16). In small telescopes it will appear as a slightly elongated hazy patch, but larger apertures reveal a rich cluster that seemingly becomes increasingly resolvable the longer you view it (see Figure 3.20).

Then there is **NGC 103**, which will appear as a smudge of light in small telescopes (see Star Chart 3.16), but will reveal itself as a nice group of about 20 stars of magnitude 11 and 12 (see Figure 3.21).

---

[3] It always makes me wonder who catalogues these objects and decides when a group of stars is a cluster or just a pleasing arrangement. Surely five stars isn't a cluster?

Figure 3.18 (*above*). Messier 52 (Klaus Eder and Georg Emrich, AAS Gahberg).
Figure 3.19 (*below*). NGC 7789 (Klaus Eder and Georg Emrich, AAS Gahberg).

Figure 3.20. NGC 7790 (Harald Strauss, AAS Gahberg).

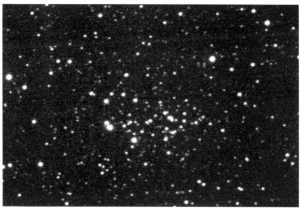

A cluster that has a wedge shape is **NGC 129 (Herschel 78)** (see Star Chart 3.16). This is a bright, open cluster, irregularly scattered and uncompressed, making it difficult to distinguish from the background, and it seems to be in two groups. One is an arc of stars to the northeast, and the other is a small asterism resembling a V shape (see Figure 3.22). Up to a dozen stars can be seen with binoculars, but many more are visible under telescopic aperture. Under good observing conditions and using averted vision, the unresolved background stars of the cluster can be seen as a faint glow.

A bit more of a challenge is **NGC 136 (Herschel 35)**. This is a very small cluster, that actually looks like a tiny sprinkling of diamond dust. Although it can be observed with a 15 cm telescope, it needs a very large aperture of at least 20 cm to be fully resolvable. It always seems to me to have a faint hazy background glow of unresolvable stars (see Figure 3.23). One cluster that is often overlooked is **King 14**. This cluster is a faint but rich object. With a 10 cm aperture telescope, several stars can be resolved set against a faint glow.

Two clusters that are faint and small, but nevertheless are worth searching out, are **NGC 146** and **NGC 189**. Both are small and irregularly shaped, and have about 15 to 20 cluster members. The former (see Figure 3.24) can be seen in small telescopes of aperture 10 cm, whereas the latter (see Figure 3.25) is best seen at a slightly larger aperture. It goes without saying that the larger the aperture the more stars you will see in these two objects.

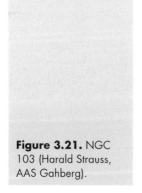

Figure 3.21. NGC 103 (Harald Strauss, AAS Gahberg).

Figure 3.22. NGC 129 (Space Telescope Science Institute, AAO, UK–PPARC, ROE, National Geographic Society, and California Institute of Technology).

A cluster that is also associated with some nebulosity is **NGC 281 (IC 1590)** (see Star Chart 3.17). Both objects have the same designation and in a medium-aperture telescope the cluster appears as a collection of nearly 30 stars of 7–11th magnitude. The nebulosity can be seen as a very slight haze, but in larger telescopes with an appropriate filter it is very apparent, nearly 1° in size. Some people see it as a maple leaf shape, but I think it resembles the North American Nebula in some respects (see Figure 3.26). What do you see?

Figure 3.23. NGC 136 (Space Telescope Science Institute, AAO, UK–PPARC, ROE, National Geographic Society, and California Institute of Technology).

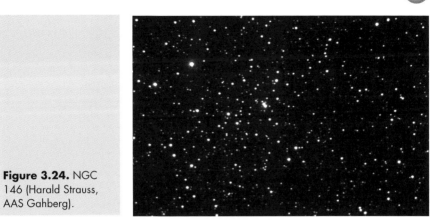

Figure 3.24. NGC 146 (Harald Strauss, AAS Gahberg).

A faint cluster that is also rich and compressed is **NGC 381 (Herschel 64)** (see Star Chart 3.18). It can be resolved with an aperture of 10 cm, but with medium aperture of, say, 20–25 cm, over 60 stars of 12th and 13th magnitude become visible (see Figure 3.27). **NGC 433 (Stock 22)** is a small and compact cluster of about 20 stars of magnitude 11–15 that are located around a 9th magnitude star.

Three clusters that make fine targets for small telescopes of aperture 10–20 cm are **NGC 436 (Herschel 45)** (see Star Chart 3.18), **NGC 457 (Caldwell 13, Herschel 42)** and **NGC 559 (Herschel 48, Caldwell 8)**. The first is a faint and small cluster that is nevertheless fairly rich. It is readily seen even though it is small and contains many 11th and 12th magnitude stars set against the barely resolved remaining cluster stars (see Figure 3.28). The second cluster, Caldwell 13, also known as the **Owl Cluster**, is a wonderful cluster, and can be considered one of the finest in Cassiopeia. It is easily seen in binoculars as two

Figure 3.25. NGC 189 (Space Telescope Science Institute, AAO, UK–PPARC, ROE, National Geographic Society, and California Institute of Technology).

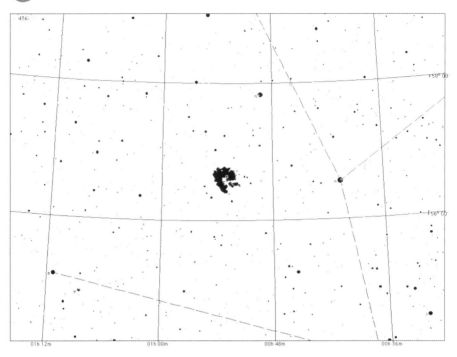

Star Chart 3.17 (*above*). NGC 281.
Star Chart 3.18 (*below*). NGC 436; NGC 381; NGC 581; NGC 659; NGC 654; NGC 663; Messier 103; NGC 457.

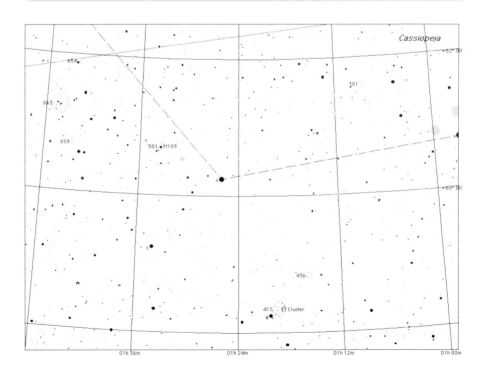

Figure 3.26. NGC 281 (Klaus Eder and Georg Emrich, AAS Gahberg).

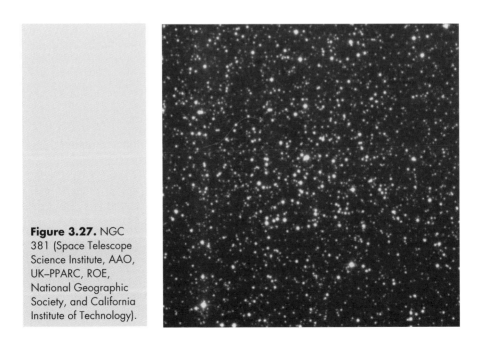

Figure 3.27. NGC 381 (Space Telescope Science Institute, AAO, UK–PPARC, ROE, National Geographic Society, and California Institute of Technology).

Figure 3.28. NGC 436 (Harald Strauss, AAS Gahberg).

southward-arcing chains of stars, surrounded by many fainter components (see Figure 3.29). The gorgeous blue and yellow double **Phi (φ) Cassiopeiae**, and a lovely red star, **HD 7902**, lie within the cluster. Located at a distance of about 8000 light years, this young cluster is located within the **Perseus Spiral Arm** of our Galaxy. The third cluster is a small, faint object that will appear literally as a patch of stardust (see Figure 3.30). There are several 11th and 12th magnitude stars present and many more just beyond the limit of resolution.

Another Messier object is **Messier 103 (NGC 581)** (see Star Chart 3.18). This is a nice rich cluster of stars, which is resolvable in small binoculars as a group of around thirty 10th magnitude and fainter stars in an area of only 6 arcminutes. Using progressively larger apertures, more and more of the cluster will be revealed (as with most clusters). It has a distinct fan shape, and the star at the top of the fan is **Σ (Struve) 131**, a double star with colors reported as pale yellow and blue (see Figure 3.31). Close by is also a rather nice, pale, red-tinted star.

The cluster is the last object in Messier's original catalogue and lies in an area that is rich on open star clusters including **NGC 663 (Caldwell 10)**, **NGC 654** and **NGC 659** (see Star Chart 3.18). These three clusters are all visible in telescope of 10 cm. They are small and faint, but rather rich and make ideal objects to view. With NGC 659, it seems as if the cluster is just an enhancement of the rich Milky Way background (see Figure 3.32), while in NGC 663, one of the constellations unknown delights, a dark dust lane can be seen edging into the cluster from the northwest (see Figure 3.33).

Figure 3.29. NGC 457 (Klaus Eder and Georg Emrich, AAS Gahberg).

Figure 3.30. NGC 559 (Harald Strauss, AAS Gahberg).

A definite challenge to observers, and perhaps the reason why it is nearly unknown to most, is **Trumpler 1 (Collinder 15)**. Even with a telescope of 12 cm aperture, this small and tightly compressed cluster will be a challenge. However, in larger apertures, it is a delightful object of nearly 30 stars of 10–14th magnitude.

Two clusters that should be on everybody's observing schedule are **Stock 2** and **Stock 5**. Stock 2 is another undiscovered and passed-over cluster! It is wonderful in binoculars and small telescopes and lies 2° north of its more famous cousin the Double Cluster. At nearly a degree across it contains over fifty 8th magnitude and fainter stars. Well worth seeking out. Stock 5 can be seen with the naked eye as a faint spot set in the Milky Way. It lies southwest of three bright stars, one of which is **52 Cassiopeiae**. Note that the cataloged coordinates seem to be in error here, as at the stated position no cluster is seen. The cluster in question lies just a little south as a circlet of stars.

Another cluster that has some associated nebulosity is **Melotte 15** and **IC 1805,** respectively (see Star Chart 3.19). In a telescope as small as 6 cm, the cluster will appear as a loose scattering of over 20 stars of 8th magnitude and fainter. The nebulosity however only becomes visible in telescopes of aperture 25 cm and more, and even then a filter will be needed.

One cluster that seems to me to be more of an enrichment of the Milky Way is **NGC 1027** (**Herschel 66**) (see Star Chart 3.19). It is a bright and rich cluster but rather scattered (see Figure 3.34).

Figure 3.31. Messier 103 (Harald Strauss, AAS Gahberg).

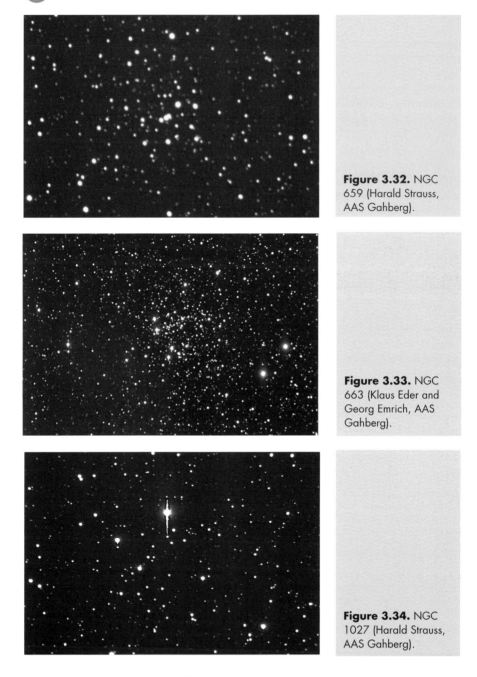

Figure 3.32. NGC 659 (Harald Strauss, AAS Gahberg).

Figure 3.33. NGC 663 (Klaus Eder and Georg Emrich, AAS Gahberg).

Figure 3.34. NGC 1027 (Harald Strauss, AAS Gahberg).

Our final open clusters are **Collinder 33** and **Collinder 34**. Both are large and scattered groupings of stars and in fact in telescopes of all apertures they are difficult to make out from the surrounding rich star fields. Collinder 33 is the slightly brighter of the two, and in the largest of telescopes, say aperture 30 cm or more, a faint nebulosity can be seen surrounding both.

Figure 3.35. NGC 7635 (Harald Strauss, AAS Gahberg).

That concludes our tour of the open clusters in Cassiopeia. There are many more readily visible to owners of large telescopes, but I will leave those to the more detailed and object-specific observing guides.

The constellation has a surprising amount of nebulosity that has been photographed or CCD imaged to great effect, but alas much of this is not within the reach of small to medium-sized amateur telescopes. There are a few patches that are readily seen under the right conditions, but for the most part they are elusive and dim. Nevertheless, I will mention the brighter ones here in the hope that some of you will be able to see them.

In addition to the nebulae I mentioned above associated with star clusters, there are a few others we can see. Our first nebula is **NGC 7635** (**Caldwell 11**). This nebula, also known as the **Bubble Nebula**, is very faint, even in telescopes of aperture 20 cm, as an 8th magnitude star within the emission nebula and a nearby 7th magnitude star hinder its detection owing to their combined glare. The use of averted vision will help in its detection, however (see Figure 3.35). Furthermore, it appears that a light filter does not really help here. Research suggests that a strong stellar wind from a star pushes material out – the "Bubble" – and also heats up a nearby molecular cloud, which in turn ionizes the "Bubble". It really does bear a striking resemblance to a soap bubble.

Two other emission nebulae that are very faint are **IC 59** and **IC 63**, which are really a melange of both emission and reflection nebulae (see Figure 3.36). They are near to Gamma (γ) Cassiopeiae (see later) which thus interferes with their detection. Both are fairly large and IC 63 has a brighter area on its southern rim. The use of averted vision and a UHC filter will be needed without a doubt. Also, it may be an idea to use an occulting bar to block out the light from the aforementioned star. Near to Melotte 15 are **NGC 896** (Star Chart 3.19) and **IC 1795**, both emission nebulae. Using a large telescope, an [OIII] filter and a very dark sky, it should be no problem to see these objects. In fact, there seems to be

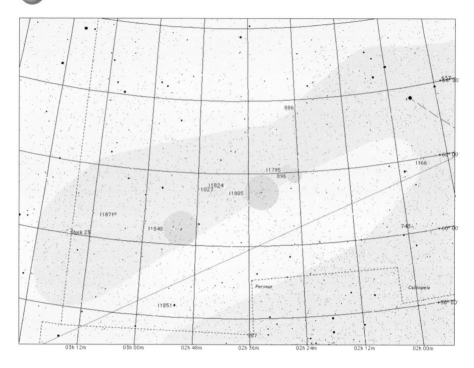

Star Chart 3.19. IC 1027; IC 1805; IC 1848; IC 1785; NGC 896.

Figure 3.36. IC 63 (Space Telescope Science Institute, AAO, UK–PPARC, ROE, National Geographic Society, and California Institute of Technology).

Figure 3.37. NGC 896 (Klaus Eder and Georg Emrich, AAS Gahberg).

more nebulosity in this area of the sky than is often indicated on star charts (see Figure 3.37). However, both will benefit from the use of averted vision.

Surrounding the star cluster **NGC 1848**, and having the same designation, is a very large and diffuse emission nebula of oval shape (see Star Chart 3.19). It has a low surface brightness and so needs the best of conditions and equipment. Many observers report that the nebulosity seems to be concentrated around individual stars of the cluster, but this may just be a case of the stars illuminating the nebulosity.

Considering that the constellation is swathed in the nebulosity and star fields of the Milky Way, it should come as a surprise to know that there are a few galaxies that manage to peek out at us. Firstly there is **NGC 147 (Caldwell 17)** (see Star Chart 3.20). It is difficult to locate and observe, however, so dark skies are a prerequisite. It has been said that a minimum of 20 cm aperture is needed to see this galaxy, but I have recently had reports that under excellent conditions a 10 cm telescope is sufficient, though averted vision was needed (see Figure 3.38). The moral of this story is that dark skies are essential to see faint objects. Increased aperture will help, as well as higher magnification, when its nuclear region then becomes visible. The galaxy is classified as a dwarf elliptical galaxy but what is surprising is that although some distance from **Messier 31**, the **Andromeda Galaxy,** it is in fact a companion to that famous galaxy. A member of the **Local Group**, it is one of over 30 galaxies which are believed to be companions to either M31 or the Milky Way. It shines at 9th magnitude and is some 13×8 arcminutes in size.

Another galaxy that lies close by is **NGC 185 (Caldwell 18)** (see Star Chart 3.20). This is another companion galaxy to M31. However, this is much easier to locate and observe (see Figure 3.39). In a telescope of 10 cm it will just be detected, whereas in 20 cm it is easily seen. It remains featureless even at larger apertures (40 cm), but with a perceptibly brighter core, and is around 11.5×9.8 arcminutes in size. Several reports mention that

Figure 3.38. NGC 147 (Harald Strauss, AAS Gahberg).

with the very large aperture of 75 cm some resolution of the galaxy becomes apparent. It is another dwarf elliptical galaxy.

About 3° southeast of NGC 185 is the compact elliptical galaxy **NGC 278** (see Star Chart 3.20). This is a faint 11th magnitude object some 2.2 × 2.1 arcminutes (see Figure 3.40). What makes all of these galaxies nice observing targets is that they can be seen with telescopes as small as 10 cm to 15 cm aperture and will appear as ghostly patches of pale grey light.

Star Chart 3.20. NGC 147; NGC 185; NGC 278.

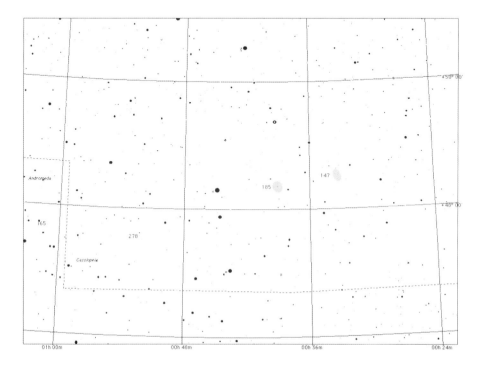

Figure 3.39. NGC 185 (Harald Strauss, AAS Gahberg).

Planetary nebulae also make an appearance in this region of the sky, but either due to an intrinsic low luminosity or obscuration by dust, they are all faint and so will need telescopes of apertures 30 cm or greater in order to be located. They are not an inspiring bunch of objects but should still get a mention. They are **Abell 84 (PK 112–10.1)**, **PK 119–6.1**, **IC 1747 (PK 130 +1.1)** and **IC 289** (see Star Chart 3.21). The first is a very faint 13th magnitude object that will require an [OIII] filter to be seen, whereas the second will appear stellar-like even with a large telescope if a low power is used. Therefore a high magnification is suggested as well as an [OIII] filter. The third planetary is only 30 arcseconds southeast of the bright star **Epsilon (ε) Cassiopeiae** and is bright[4] and round (see Figure 3.41). The fourth and final planetary is a pale, round object that really needs good conditions in order to be located.

Finally, let's not forget that there are many splendid stars in Cassiopeia; doubles, triple and variables, and so we shall have a look at just a few of them. Our first star is **Gamma (γ) Cassiopeiae**, a peculiar star having bright emission lines in its spectrum, indicating that it ejects material in periodic outbursts. It varies in magnitude from about 1.5 to 3. The middle star of the familiar W shape of Cassiopeia, it also has areas of nebulosity around it (see above). It is the prototype of a class of irregular variable star that are believed to be rapidly spinning. A fine double star is **Sigma (σ) Cassiopeiae**. Located within a splendid

Figure 3.40. NGC 278 (Harald Strauss, AAS Gahberg).

---

[4] By bright I mean bright as seen in a large telescope!

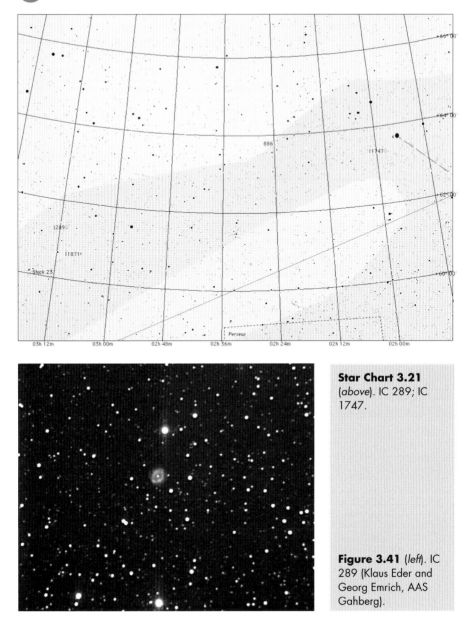

Star Chart 3.21 (*above*). IC 289; IC 1747.

Figure 3.41 (*left*). IC 289 (Klaus Eder and Georg Emrich, AAS Gahberg).

star field, this is a lovely bluish and yellow double system. Some observers describe the colors as green and blue. Another lovely double is **Eta (η) Cassiopeiae**. Discovered by William Herschel in 1779, this is another system which has had differing colors reported. The primary has been described as gold, yellow and topaz, while the secondary has been called orange, red and purple and has an apparently near-circular orbit. The separation varies from between 5 to 16 arcseconds over about 500 years.

Our final star, and indeed final object in this chapter, is the multiple star **Alpha (α) Cassiopeiae** and is an easy object for small telescopes. The primary has an orange tint that contrasts nicely with the bluish companion stars. The other two companions are much fainter at 12th and 13th magnitude.

The following constellations are also visible during these months at different times throughout the night. Remember that they may be low down and so diminished by the effects of the atmosphere. Also, you may have to observe them either earlier than midnight, or some considerable time after midnight, in order to view them.

Northern Hemisphere

Andromeda, Aquila, Camelopardalis, Cassiopeia, Cepheus, Sagittarius, Cygnus, Delphinus, Gemini, Hercules, Lacerta, Libra, Lupus, Lyra, Orion, Ophiuchus, Perseus, Sagitta, Sagittarius, Scorpius, Scutum, Serpens Cauda, Taurus, Vulpecula.

Southern Hemisphere

Apus, Aquila, Ara, Canis Major, Canis Minor, Carina, Chameleon, Corona Australis, Crux, Cygnus, Delphinus, Gemini, Hercules, Lacerta, Libra, Lupus, Lyra, Musca, Norma, Ophiuchus, Orion, Pavo, Puppis, Pyxis, Sagitta, Scorpius, Scutum, Serpens Cauda, Triangulum Australe, Volans, Vulpecula.

Objects in Cepheus

Stars

Designation	Alternate name	Vis. mag	RA	Dec.	Description
Mu (μ) Cephei	Garnet Star	3.4–5.1	$21^h43.5^m$	+53° 47'	Variable star
Beta (β) Cephei	8 Cephei	3.2, 7.9	$21^h28.7^m$	+70° 34'	PA 249°; Sep. 13.3"
OΣ 440		6.4, 10.7	$21^h27.4^m$	+59° 45'	PA 181°; Sep. 11.4"
Xi (ξ) Cephei	17 Cephei	4.4, 6.5	$22^h03.8^m$	+64° 38'	PA 277°; Sep. 7.7"
Krueger 60		9.8, 11.3	$22^h28.1^m$	+57° 27'	PA 137°; Sep. 3.3"
Delta (δ) Cephei	27 Cephei	3.48, 4.37	$22^h29.2^m$	+58° 25'	Variable star
U Cephei		6.7–9.2	$01^h02.3^m$	+81° 53'	Variable star

Deep-Sky Objects

Designation	Alternate name	Vis. mag	RA	Dec.	Description
NGC 6939		7.8	$20^h31.4^m$	+60° 38'	Open cluster
NGC 6946		8.8	$20^h34.8^m$	+60° 09'	Galaxy
IC 1396		3.5	$21^h39.1^m$	+57° 30'	Open cluster & emission nebula
NGC 7142	Herschel 66	9.3	$21^h45.9^m$	+65° 48'	Open cluster
NGC 7160		6.1	$21^h53.7^m$	+62° 36'	Open cluster
NGC 7235		7.7	$22^h12.6^m$	+57° 17'	Open cluster
NGC 7261		8.4	$22^h20.4^m$	+58° 05'	Open cluster
NGC 7281		–	$22^h24.7^m$	+57° 50'	Open cluster/asterism
King 10		–	$22^h54.9^m$	+59° 10'	Open cluster
NGC 7510		7.9	$23^h11.5^m$	+60° 34'	Open cluster
Markarian 50		8.5	$23^h15.3^m$	+60° 28'	Open cluster
Barnard 169/169	LDN 1151	–	$21^h58.9^m$	+58° 47'	Dark nebula
Barnard 171	LDN 1151	–	$21^h03.5^m$	+58° 52'	Dark nebula
Barnard 173	LDN 1164	–	$22^h07.4^m$	+59° 10'	Dark nebula
Barnard 174	LDN 1164	–	$22^h07.3^m$	+59° 05'	Dark nebula

Designation	Alternate name	Vis. mag	RA	Dec.	Description
Harrington 11	Cep OB2	8.5	21^{h}48^m	+61°	Open cluster/stellar association
NGC 40	Caldwell 2	12.4	00^{h}13.0^m	+72° 32'	Planetary nebula
NGC 7354	PK107+2.1	12.2	22^{h}40.4^m	+61° 17'	Planetary nebula
NGC 7139	PK 104.7	13.3	21^{h}45.9^m	+63° 49'	Planetary nebula

Objects in Andromeda

Stars

Designation	Alternate name	Vis. mag	RA	Dec.	Description
Groombridge 34		8.2, 10.6	21^{h}28.7^m	+00° 17.9'	PA 62°; Sep. 40.0"
Gamma (γ) Andromedæ		2.3, 5.5	02^{h}03.9^m	+42° 19'	PA 63°; Sep. 9.8"

Deep-Sky Objects

Designation	Alternate name	Vis. mag	RA	Dec.	Description
NGC 7686	Herschel 69	5.6	23^{h}30.2^m	+49° 08'	Open cluster
NGC 7662	Caldwell 22/Blue Snowball	8.3	23^{h}25.9^m	+42° 33'	Planetary nebula
NGC 7640	Herschel 6000	11.3	23^{h}22.1^m	+40° 51'	Galaxy

Objects in Camelopardalis

Stars

Designation	Alternate name	Vis. mag	RA	Dec.	Description
Σ 390	Struve 390	5.1, 9.5	03ʰ 30.0 ᵐ	+55° 27'	PA 159°; Sep. 14.8"
Σ 485	SZ	7.0, 7.1	04ʰ 07.9 ᵐ	+62° 20'	PA 304°; Sep. 17.9"
Σ 550	1 Camelopardalis	5.7, 6.8	03ʰ 32.0 ᵐ	+53° 55'	PA 308°; Sep. 10.3"
SZ Camelopardalis		7.0–7.29	04ʰ 07.9 ᵐ	+62° 20'	Variable star

Deep-Sky Objects

Designation	Alternate name	Vis. mag	RA	Dec.	Description
Stock 23		5.6	03ʰ 16.3 ᵐ	+60° 02'	Open cluster
Tombaugh 5		8.4	03ʰ 47.8 ᵐ	+59° 03'	Open cluster
NGC 1502		5.7	04ʰ 07.7 ᵐ	+62° 20'	Open cluster
NGC 1501	Herschel 47	11.5	04ʰ 07.7 ᵐ	+60° 55'	Planetary nebula
van den Bergh 14	Oyster Nebula	–	03ʰ 29.2 ᵐ	+59° 57'	Reflection nebula
van den Bergh 15		–	03ʰ 30.1 ᵐ	+58° 54'	Reflection nebula

Objects in Cassiopeia

Stars

Designation	Alternate name	Vis. mag	RA	Dec.	Description
Gamma (γ) Cassiopeiae		1.5–3.0	23ʰ 56.7 ᵐ	+60° 43'	Variable star
Sigma (σ) Cassiopeiae		5.0, 7.1	23ʰ 59.0 ᵐ	+55° 45'	PA 326°; Sep. 3.0"
Eta (η) Cassiopeiae		3.4, 7.5	00ʰ 49.1 ᵐ	+57° 49'	PA 280°; Sep. 64.4"
Alpha (α) Cassiopeiae		2.2, 8.9	00ʰ 40.5 ᵐ	+56° 32'	PA 280°; Sep. 64.4"

Deep-Sky Objects

Designation	Alternate name	Vis. mag	RA	Dec.	Description
NGC 7654	Messier 52	6.9	23^{h}24.2^m	+61° 35'	Open cluster
King 12		9.0	23^{h}53.0^m	+61° 58'	Open cluster
Harvard 21		9.0p	23^{h}54.1^m	+61° 46'	Open cluster
NGC 133	Collinder3	9.4p	00^{h}31.2^m	+63° 22'	Open cluster
NGC 7789	Herschel 30	6.7	23^{h}57.0^m	+56° 44'	Open cluster
NGC 7790	Herschel 56	8.5	23^{h}58.4^m	+61° 13'	Open cluster
NGC 103		9.8	00^{h}25.3^m	+61° 21'	Open cluster
NGC 129	Herschel 78	6.5	00^{h}29.9^m	+60° 14'	Open cluster
NGC 136	Herschel 35	–	00^{h}31.5^m	+61° 32'	Open cluster
King 14		8.5	00^{h}31.9^m	+63° 10'	Open cluster
NGC 146		9.1	00^{h}31.1^m	+63° 18'	Open cluster
NGC 189	Herschel 707	8.8	00^{h}39.6^m	+13° 04'	Open cluster/emission nebula
NGC 281	IC 1590	7.4p	00^{h}52.8^m	+56° 37'	Open cluster
NGC 381	Herschel 64	9.3p	01^{h}08.3^m	+61° 35'	Open cluster
NGC 433	Stock 22	–	01^{h}15.3^m	+60° 08'	Open cluster
NGC 436	Herschel 45	8.8	01^{h}15.6^m	+58° 49'	Open cluster
NGC 457	Caldwell 13/Owl Cluster	6.4	01^{h}19.1^m	+58° 20'	Open cluster
NGC 559	Caldwell 8	9.5	01^{h}29.5^m	+63° 18'	Open cluster
NGC 581	Messier 103	7.4	01^{h}33.2^m	+60° 42'	Open cluster
NGC 663	Caldwell 10	7.1	01^{h}46.0^m	+61° 15'	Open cluster
NGC 654	Herschel 46	6.5	01^{h}44.1^m	+61° 53'	Open cluster
NGC 659	Herschel 65	7.9	01^{h}44.2^m	+60° 42'	Open cluster
Trumpler 1	Collinder 15	8.1	01^{h}35.7^m	+61° 17'	Open cluster
Stock 2		4.4	02^{h}15.0^m	+59° 16'	Open cluster
Stock 5		–	02^{h}04.5^m	+64° 26'	Open cluster
Melotte 15		6.5	02^{h}32.7^m	+61° 27'	Open cluster
IC 1805			02^{h}33.4^m	+61° 26'	Emission nebula
NGC 1027	Herschel 66	6.7	02^{h}42.7^m	+61° 33'	Open cluster

Objects in Cassiopeia (*continued*)

Designation	Alternate name	Vis. mag	RA	Dec.	Description
Collinder 33		5.9p	02^h 59.3^m	+60° 24'	Open cluster
Collinder 34		6.8p	03^h 00.9^m	+60° 25'	Open cluster
NGC 7635	Caldwell 11/Bubble nebula	–	23^h 20.7^m	+61° 12'	Emission nebula
IC 59		–	00^h 56.7^m	+61° 04'	Emission & reflection nebula
IC 63		–	00^h 59.5^m	+60° 49'	Emission & reflection nebula
NGC 896		–	02^h 24.8^m	+61° 54'	Emission nebula
IC 1795		–	02^h 26.5^m	+62° 04'	Emission nebula
NGC 1848		6.5	02^h 51.2^m	+60° 26'	Open cluster
NGC 147	Caldwell 17	9.5	00^h 33.2^m	+48° 30'	Galaxy
NGC 185	Caldwell 18	9.2	00^h 39.0^m	+48° 20'	Galaxy
NGC 278	Herschel 159	10.8	00^h 52.1^m	+47° 33'	Galaxy
Abell 84	PK 112–10.1	13.0	23^h 47.7^m	+51° 24'	Planetary nebula
PK 119–6.1		12.3	00^h 28.3^m	+55° 58'	Planetary nebula
IC 1747	PK 130 +1.1	12.1	01^h 57.6^m	+63° 20'	Planetary nebula
IC 289		13.3	03^h 10.3^m	+61° 19'	Planetary nebula

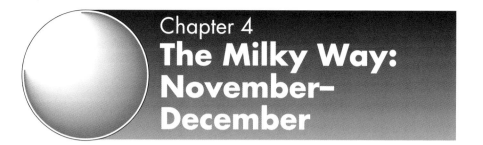

Chapter 4
The Milky Way: November–December

Perseus, Auriga, Taurus, Gemini, Orion
RA 1ʰ to 6ʰ; Dec. 60º to 10º; Galactic longitude[1] 130º to 215º; Star Chart 4

4.1 Perseus

As autumn turns to winter for northern observers we are still in those parts of the sky that are rich in Milky Way objects. However, some of the constellations are getting very low during the southern summer months and may prove difficult for southern telescopes (see Figure 4.1).

The galactic equator runs through our first constellation, **Perseus**, and the area is rich in star fields, clusters and some nebulae that are perfect for observing with binoculars (see Star Chart 4.1). It also has, surprisingly for such a rich, and one presumes, dust-filled region, several galaxies. The constellation transits in early November. But we shall start with something different, an object that is truly vast.

Our first object is the cluster called **Melotte 20**, or rather, the **Alpha Persei Stream (Perseus OB–3)**. This is a group of about 100 stars including, **Alpha (α) Persei, Psi (ψ) Persei, 29** and **34 Persei**. The stars **Delta (δ)** and **Epsilon (ε) Persei** are believed to be amongst its most outlying members, as they also share the same space motion as the main groups of stars.[2] The inner region of the stream is measured to be over 33 light years in length – the distance between 29 and ψ Persei. The group covers almost 3° and in binoculars and small telescopes the sight is a stunning panorama of jewel-like stars that are seemingly grouped into small clusters and asterisms. Oddly enough the group has neither a Messier nor an NGC designation and this tends to be overlooked. Let's remedy that now. Set against the backdrop of the irresolvable Milky Way, this is a stellar showpiece of the northern sky.

Another showpiece of the sky is the wonderful **Double Cluster** in Perseus: **NGC 869** and **NGC 884, h Persei** and **Chi (χ) Persei (Caldwell 14)**. This is a glorious object and should be on every amateur's observing schedule as it is a highlight of the northern hemisphere winter sky (see Star Chart 4.2). Strangely, it was never catalogued by Messier even though

---

[1] See Appendix 1 for details on astronomical coordinate systems.
[2] The bright stars that extend from Perseus, Taurus and Orion, and down to Centaurus and Scorpius, including the Orion and Scorpius–Centaurus associations, lie at an angle of about 1.5° to the Milky Way, and thus to the equatorial plane of the Galaxy. This group or band of stars is often called **Gould's Belt**.

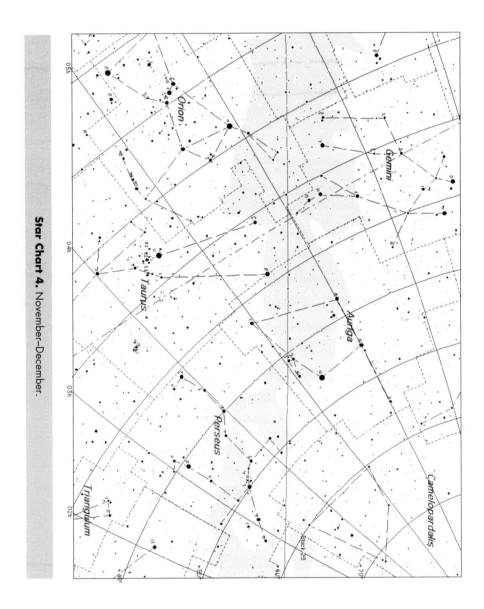

Star Chart 4. November–December.

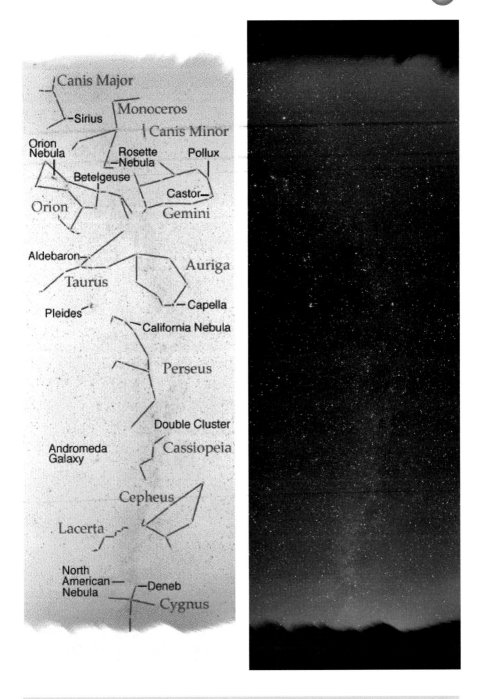

Figure 4.1. The winter Milky Way (Thor Olson)

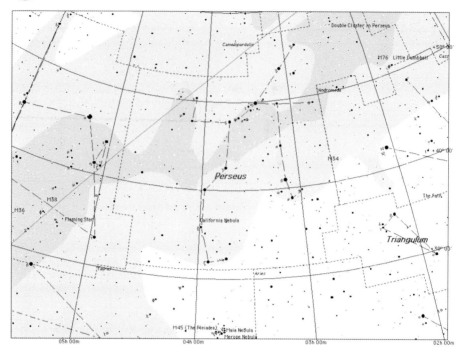

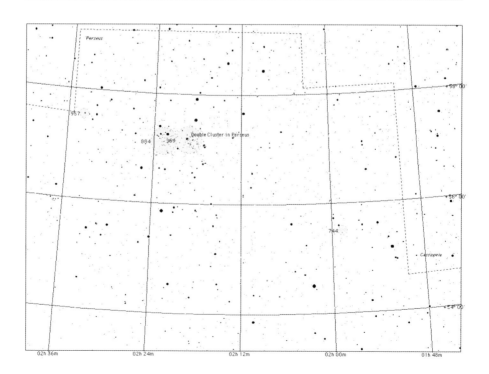

Star Chart 4.1 (*above*). Perseus.
Star Chart 4.2 (*below*). NGC 884; NGC 869; NGC 744.

Figure 4.2. NGC 869; NGC 884 (Klaus Eder and Georg Emrich, AAS Gahberg).

it is visible to the naked eye as a faint blur of light about halfway between the tip of Perseus and the "W" shape of Cassiopeia. However, it is best seen using a low-power, wide-field optical system (see Figure 4.2). But whatever system is used, the views are marvelous. NGC 869 has around 200 members, while NGC 884 has about 150. Both are composed of A-type and B-type supergiant stars with many nice red giant stars. However, the systems are dissimilar: NGC 869 is 5.6 million years old (at a distance of 7200 light years), whereas NGC 884 is younger at 3.2 million (at a distance of 7500 light years). But be advised that in astrophysics, especially in distance and age determination, there are very large errors! The Double Cluster and its surrounding Perseus Association are responsible for the name of the spiral arm that is the next spiral arm out from our own Orion–Cygnus Arm, and is called the **Perseus Spiral Arm**.

There are even more great open clusters to look at, including **Messier 34 (NGC 1039)**. This is a nice cluster easily located, and is about the same size as the full moon (see Star Chart 4.3). It can be glimpsed with the naked eye but is best seen with medium-sized binoculars, as a telescope will spread out the cluster and so lessen its impact (see Figure 4.3). There are around 70 stars shining at 7–13th magnitude. At the center of the cluster is the double star **H1123**, both members being 8th magnitude, and of type A0. The pure-white stars are very concentrated toward the cluster's center, while the fainter members disperse toward its periphery. The cluster is thought to be about 200 million years old, lying at a distance of 1500 light years.

Then there is **Trumpler 2**, an open cluster some 2° west of Eta Persei. There are about twenty 7th magnitude white and blue stars set amongst fainter members. Another small cluster is **NGC 744**, which consists of about 20 or so 10th magnitude and fainter stars (see Star Chart 4.2). It is spread over an area of 7 arcminutes and can easily be seen in a telescope of aperture 15 cm (see Figure 4.4).

Faint clusters abound in Perseus, and prime examples are NGC 1220 and King 5. The first, **NGC 1220** is a faint and small object that consists of about twelve 7–13th magnitude

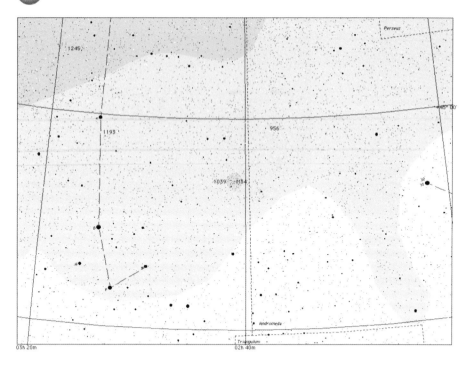

Star Chart 4.3 (*above*).Messier 34.
Figure 4.3 (*below*).Messier 34 (Harald Strauss, AAS Gahberg).

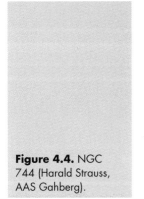

Figure 4.4. NGC 744 (Harald Strauss, AAS Gahberg).

stars (see Figure 4.5 and Star Chart 4.4), while **King 5** is very similar with half a dozen stars set against a pale haze of unresolved stars. Both clusters are visible in telescopes of 15 cm or more.

Several clusters are perfect for small telescopes and we shall look at a few of these now. One such cluster is **NGC 1245 (Herschel 25)** (see Star Chart 4.5). Even in a telescope as small as 10 cm aperture, it will show as a round and hazy, albeit faint patch that is nicely framed by a triangle of 8–10th magnitude stars (see Figure 4.6). With larger apertures of, say, 20 cm, many more stars become visible and in the largest telescopes the cluster becomes a very nice object indeed.

Following in a similar vein is **NGC 1342 (Herschel 88)** also visible in a 10 cm aperture telescope, but this time it is a bright and large cluster (see Figure 4.7). Once again it improves when larger apertures are used (see Star Chart 4.6).

A faint but rich cluster that some observers have reported as looking like a figure "9" is **NGC 1513 (Herschel 60)**. Visible in small telescopes, it will just appear as a hazy patch, but gradually becomes more impressive the larger the aperture (see Figure 4.8).

Our final two clusters are also fine targets for small telescopes: **NGC 1528 (Herschel 61)** and **NGC 1545**. The former, NGC 1528, is a large and bright cluster with about 40 members that gets decidedly more concentrated towards its central region (see Figure 4.9). There are some obvious arcs and chains of stars to be seen here, although in binoculars only a few stars can be seen whilst the remainder form a hazy background glow (see Star Chart 4.7). In the latter, NGC 1545, a small telescope will only show a faint grouping of stars that surrounds three brighter stars (see Figure 4.10). These foreground stars make the integrated magnitude for the cluster 6.2 or even higher, even though they are not really members of the cluster itself (see Star Chart 4.7). As before, using larger telescopes will improve the objects quite a bit.

We shouldn't ignore the stars here, as there are some lovely double systems and a very famous variable star. Our first double is Σ (**Struve**) **336**. This is an easy system to resolve, of orange and white stars. Notice, however, that the colors are very faint, and more of a delicate tint. Another colorful double is **Theta (θ) Persei**, which consists of a nice 4th magnitude yellow star along with a 10th magnitude blue star. Also in the same field of view is Σ (**Struve**) **304**, which is a wide pair of white and blue stars. It lies some 40 arcminutes to the east.

Two more colorful double stars are Σ (**Struve**) **268** and Σ (**Struve**) **270**. The first is a close pair of blue and white stars, whereas the second is a wide pair of pale yellow and blue stars. Both are splendid objects for a small telescope of, say, 10 cm aperture. A highlight of the

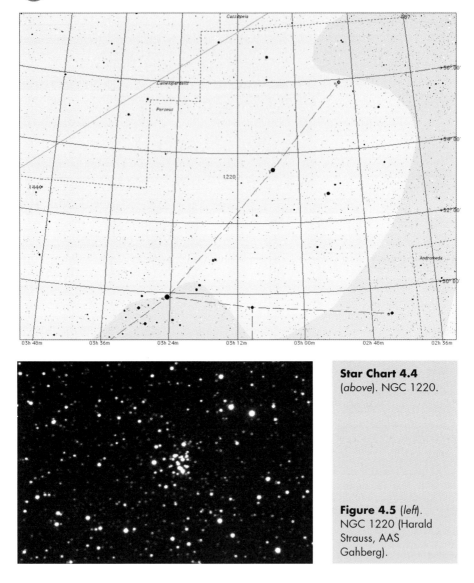

Figure 4.5 (*left*).
NGC 1220 (Harald
Strauss, AAS
Gahberg).

constellation is the glorious double star **Eta (η) Persei**. This is a lovely color-contrasted system of gold and blue stars. In addition, some 66 arcseconds to the west is a close double star of magnitudes 10 and 10.5, which, along with the few faint stars that surround the two doubles, seems to form a nice cluster. Two more doubles worthy of note are **Σ (Struve) 369** and **Σ (Struve) 392**. Both are easily resolved in small telescopes and consist of a lemon–pale blue system and a yellow and pale blue system.

There are also some multiple star systems that are observable. The first of these is **Zeta (ζ) Persei**. This consists of a bright blue-white star that has three faint companions: a 9th magnitude star some 13 arcseconds away, an 11th magnitude star some 33 arcseconds away and another 9th magnitude star nearly 100 arcseconds away. However, what makes the star even more interesting is the fact that it is the brightest member of what is called

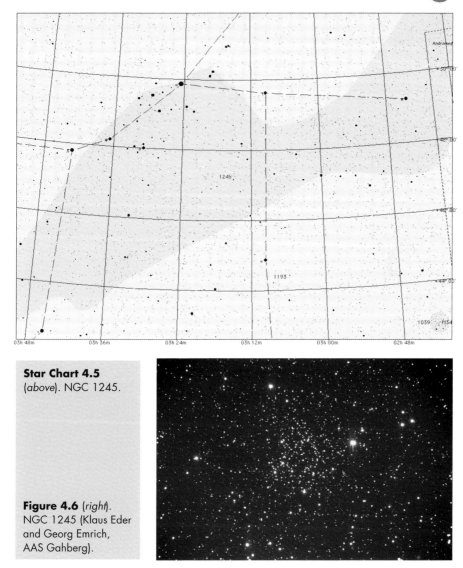

Star Chart 4.5
(*above*). NGC 1245.

Figure 4.6 (*right*).
NGC 1245 (Klaus Eder
and Georg Emrich,
AAS Gahberg).

the **Zeta Persei Association**. Also known as **Per OB2**, this association includes ζ and ξ **Persei**, as well as **40, 42** and **o Persei**. The California Nebula, NGC 1499, is also within this association. The measured expansion rate of the group suggests that the stars within it were born very recently, about 1 million years ago. Measurements reveal that the group is about 1300 light years from us. Our final multiple is **56 Persei**, which consists of a lovely golden colored primary and a pale yellowish secondary. It is set in a field that is sprinkled with background stars.

Our last star is probably the most famous in the constellation: **Algol, Beta (β) Persei**. Also known as the **Demon Star**, it is the prototype and most famous of all the eclipsing variable stars. What makes it nice for us, however, is that it is perfect for observation to all amateur astronomers even if you do not possess binoculars or a telescope. An eclipse lasts 10 hours and has been measured to an astonishing accuracy, so that we can predict with

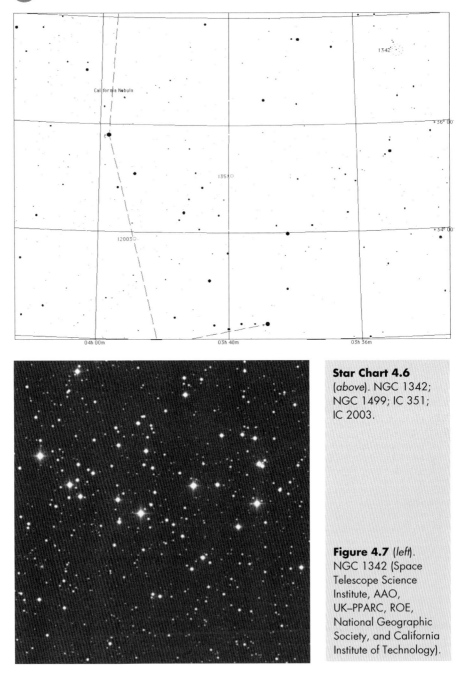

Star Chart 4.6 (*above*). NGC 1342; NGC 1499; IC 351; IC 2003.

Figure 4.7 (*left*). NGC 1342 (Space Telescope Science Institute, AAO, UK–PPARC, ROE, National Geographic Society, and California Institute of Technology).

certainty when the next cycle begins. The star undergoes an eclipse every 2.86739 days, or in terms more familiar to mere mortals, every 2 days, 20 hours, 48 minutes and 56 seconds. It takes about 5 hours for the star to fall in magnitude from 2.1 to 3.4, and so it can be observed during a single evening's observing session. On rare occasions when chance

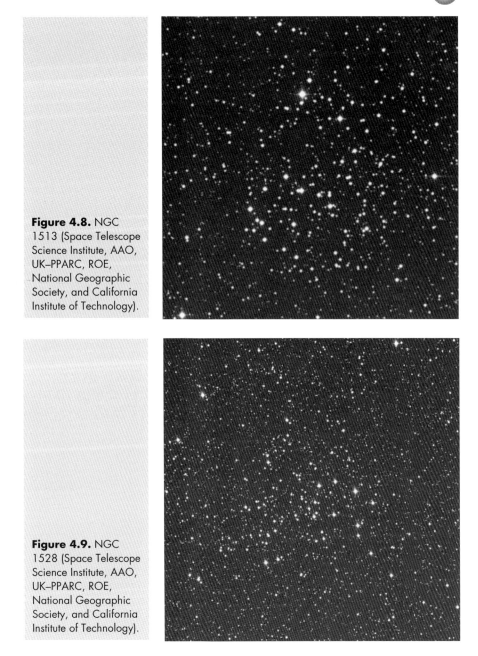

Figure 4.8. NGC 1513 (Space Telescope Science Institute, AAO, UK–PPARC, ROE, National Geographic Society, and California Institute of Technology).

Figure 4.9. NGC 1528 (Space Telescope Science Institute, AAO, UK–PPARC, ROE, National Geographic Society, and California Institute of Technology).

allows, the complete 10-hour cycle, that is from maximum to minimum back to maximum, will occur during one evening.

As this star is so famous and important[3] it may be wise if we spend a little time explaining what is actually going on here. The unresolved pair of stars comprises a cool giant star

---

[3] Not forgetting that it is perfect for observation!

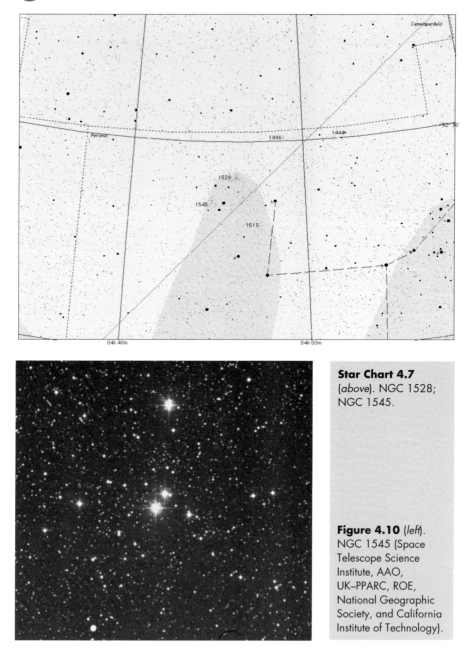

Star Chart 4.7
(*above*). NGC 1528;
NGC 1545.

Figure 4.10 (*left*).
NGC 1545 (Space
Telescope Science
Institute, AAO,
UK–PPARC, ROE,
National Geographic
Society, and California
Institute of Technology).

of type G or early K and a B8-type main-sequence star. They lie close to each other, only 6 million miles apart, and as they orbit each other the secondary covers about 79% of the primary's disk. Then about mid-way through the period a second but slight minimum occurs when the brighter star passes in front of the fainter companion. There is evidence that a third much fainter companion also exists and perhaps even a fourth. Some amateurs regard variable star observing as a pastime that takes a lot of time and practice. While this

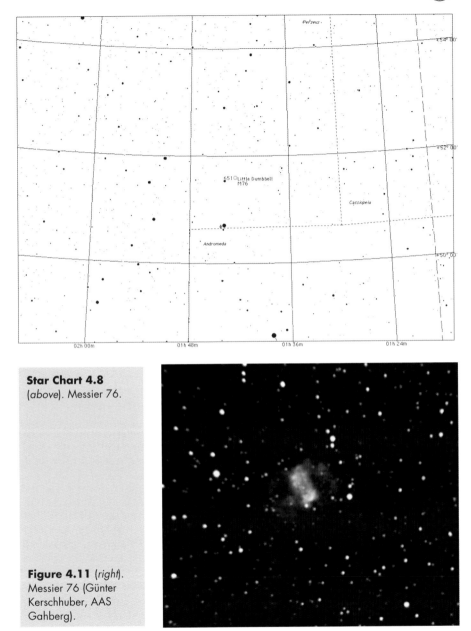

Star Chart 4.8 (*above*). Messier 76.

Figure 4.11 (*right*). Messier 76 (Günter Kerschhuber, AAS Gahberg).

may be true for the fainter and more elusive objects, Algol is a delight to observe, especially when you can actually see that the magnitude has changed during the course of your observing session.

Now let us look at some nebulae. There are some nice planetary nebulae in Perseus, and none more so than **Messier 76 (NGC 650–51)** (see Star Chart 4.8). Also known as the **Little Dumbbell Nebula** this is a small planetary nebula that shows a definite nonsymmetrical

Figure 4.12. NGC 1499 (Klaus Eder and Georg Emrich, AAS Gahberg).

shape. In small telescopes of aperture 10 cm, and using averted vision, two distinct "nodes" or protuberances can be seen (see Figure 4.11).

With apertures of around 30 cm, the planetary nebula will appear as two bright but small disks, which are in contact. Even larger telescopes will show considerably more detail, as will the use of an [OIII] filter. Another planetary we can observe that may, however, appear star-like in smaller telescopes is **IC 351 (PK 159–15.1)**. It is very small, only 7 arcseconds across, which probably explains why it is often passed over as a star. It is also faint, at 12th magnitude, which doesn't help in its identification. In very large telescopes it will appear with a nice bluish-green tint. Our final planetary is **IC 2003 (PK 161–14.1)**. In medium-aperture telescopes, say 20 cm or more, unless you use a high enough magnification, it will remain stellar in appearance. Under high magnification it appears as a disk-shaped object of magnitude 12.5. The central star is an exceedingly faint 15th magnitude and very hard to see even in big telescopes.

Another famous object we can look at is the emission nebula **NGC 1499**, often referred to as the **California Nebula**. This emission nebula presents a paradox. Some observers state that it can be glimpsed with the naked eye, others that binoculars are needed. The combined light from the emission nebula results in a magnitude of 6, but the surface brightness falls to around 14th magnitude when observed through a telescope (see Figure 4.12). Most observers agree, however, that the use of filters is necessary, maybe a Hβ filter, especially from an urban location and when the seeing is not ideal. Clean optics is also a must to locate this nebula. Glimpsed as a faint patch in binoculars, with telescopes of aperture 20 cm the emission nebula is seen to be nearly 3° long. (See Star Chart 4.6.)

Whatever optical instrument is used, this nebula will remain faint and elusive. In binoculars it will appear as a rectangular patch of pale light north of the star **46 Persei,** believed to be the star responsible for providing the energy that makes the gas glow in the nebula.

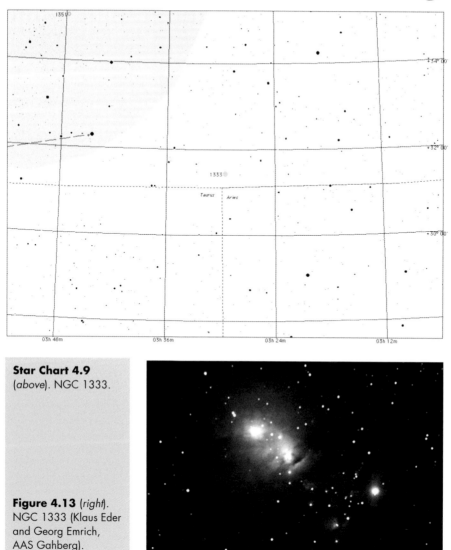

Star Chart 4.9 (*above*). NGC 1333.

Figure 4.13 (*right*). NGC 1333 (Klaus Eder and Georg Emrich, AAS Gahberg).

With a telescope of aperture 20 cm, the nebula is about 3° long, although still difficult to locate. Southern observers may find it difficult to locate, as it is rather low. It may appear fabulous in the many images one sees of it, but try finding it!

A rare reflection nebula that can also be observed is **NGC 1333**. This is a nice, easily seen reflection nebula, and appears as an elongated hazy patch (see Star Chart 4.9). Larger-aperture telescopes will show some detail along with two fainter dark nebulae **Barnard 1** and **Barnard 2**, lying toward the north and south of the reflection nebula. It can be seen with a telescope of aperture 20 cm, but as with all reflection nebulae will need a dark night and clean optics (see Figure 4.13).

One emission nebula that will need at least an aperture of 25 cm or more in order to be seen is **NGC 1491 (Herschel 258)**. It lies just to the west of an 11th magnitude star and appears quite diffuse and bright, relatively speaking, shaped like a wedge some 6

Figure 4.14. NGC 1491 (Klaus Eder and Georg Emrich, AAS Gahberg).

arcseconds across (see Figure 4.14). Larger telescopes will begin to show some considerable detail within the nebula itself.

We now come to the topic of galaxies. Let me say straight away that there are a lot of galaxies in Perseus, including a galaxy cluster known as the **Perseus I Galaxy Cluster (Abell 426)**. That was the good news. The bad news is that most are only suitable for apertures of 25–30 cm and larger. This means that they are beyond the means of most of us and we will only look at a few brighter examples.

One such galaxy is **NGC 1023 (Herschel 156)**. Considering what I said immediately above, it is ironic that this galaxy can be observed in a telescope as small as 10 cm, where it will appear as a largish elongated lenticular galaxy with a very bright and apparent center[4] (see Figure 4.15). Larger telescopes will of course show far more detail, including a bright lens-shaped halo and concentrated core (see Star Chart 4.10). The galaxy is the brightest member of a group of galaxies that includes NGC 891 in Andromeda. It is about 10 million parsecs from us.

A galaxy that needs a large aperture is **NGC 1275**. This is located near the center of the Perseus I Cluster and is a strong extragalactic radio source known as **Perseus A (3C84)** (see Star Chart 4.11). It may be that the galaxy is actually colliding with another, or merging. It is the brightest member of the Perseus Cluster and is about 300 million light years away. In

Figure 4.15. NGC 1023 (Klaus Eder and Georg Emrich, AAS Gahberg).

---

[4] It is, however, the only galaxy that can be seen in small telescopes.

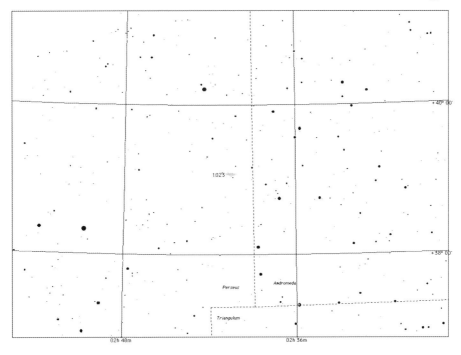

Star Chart 4.10 (*above*). NGC 1023.
Star Chart 4.11 (*below*). NGC 1275.

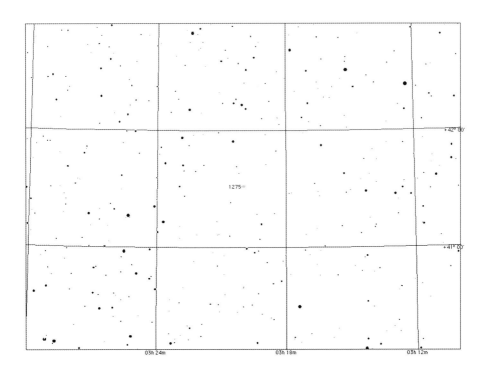

Figure 4.16. NGC 1275 (AURA/NOAO/NSF).

a large telescope it will appear as a small faint nucleus surrounded by an even fainter halo (see Figure 4.16). As an aside, there is a string of galaxies about 1° long spreading to the west and surrounding area of NGC 1275, which can also be glimpsed under perfect conditions.

4.2 Auriga

Auriga is a constellation that many northern observers recognize, and its appearance is a signal that autumn is coming to an end and winter will soon be with us, with its attendant frosts and clear nights. It is low in the sky for southern observers and in fact may be partly hidden, but that should not deter anyone from looking in detail here, as the Milky Way passes through most of the constellation, with only the northernmost reaches left untouched.

It is a worthwhile constellation to scan with binoculars as it is full of faint and surprising clusters and star fields just on the edge of visibility (see Star Chart 4.12). There is a lot here for us to observe, including many bright and colorful double stars, some lovely open clusters, and a few nebulae (see Figure 4.17). It transits in early December.

Let's begin with some nice double star systems of which there are a lot, although we will just look at a few. Our first double star is **Omega (ω) Aurigae**. These stars appear white and blue in small telescopes, but have shown subtle tints in larger instruments. Indeed some observers see a pair of yellow stars, one pale and one deeply tinted. Easily visible in telescopes as small as 5 cm. Another fine double for very small telescopes is **14 Aurigae**, which appears as a pair of lemon-yellow and blue stars, and if a large telescope is used, a third star will be seen, pale blue in color. The brighter star is in fact a small-amplitude variable star of the **Delta (δ) Scuti** type, designated **KW Aurigae**. A star that is very difficult to observe but nevertheless well worth the effort is **UV Aurigae.** It is difficult to locate owing to its variable nature. It is a carbon star, coupled with a B-type giant star. Persevere and you will be rewarded by a lovely combination of orange and blue stars. But you will need a medium to large aperture telescope in order for this lovely pair to be really appreciated.

Although our next star is difficult to split, it can be seen in a 5 cm telescope, although not clearly. **Theta (θ) Aurigae** is a star whose spectrum shows strong lines of silicon.

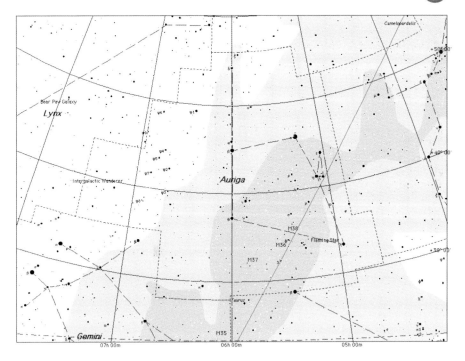

Star Chart 4.12. Auriga.

Small telescopes, excellent conditions and superb optics are required to see these two bluish-white stars and can be used as a test for a 10 cm telescope. In a larger telescope, the stars will appear as white and blue, but you may need a high magnification in order to resolve the system completely. The brighter star is another of the small-amplitude variable stars of the **Delta (δ) Scuti** type. A star that is located right at the fringes of the Milky Way is **Psi⁵ (φ⁵) Aurigae**. This is a nice pair of yellow and blue stars set against a backdrop of faint stars that is easily seen in very small telescopes.

Two more systems that are perfect for very small telescopes are **Σ (Struve) 928** and **Σ (Struve) 929**. The former is a lovely pair of near-equal-magnitude stars, while the latter is a nicely color-contrasted pair of pale-yellow and blue stars. Set in a lovely star field is the double **Σ (Struve) 698**, a pair of yellowish-orange stars easily seen in small telescopes.

There is an exquisite triple star system for us to look at: **OΣ 147**. This is a wonderful triple star system forming a triangle of yellow and blue stars. It is visible in all sizes of telescopes and you may also see, under suitable conditions, that the third component is also a very close double star itself.

A trio of famous variable stars are also visible to us: AE Aurigae, Beta (β) Aurigae and RT Aurigae. The first, **AE Aurigae** is a strange O-type star that exhibits irregular variations in magnitude and illuminates the Flaming Star Nebula, IC 405. Research indicated that the star has only recently[5] encountered the nebula, clearing a path through it as it passes. Furthermore, the star is one of three **runaway stars** that are receding from the Orion Association (see later in this chapter), the other two being **53 Arietis** and **Mu (μ) Columbae**.

---

[5] Astronomically speaking, of course!

Figure 4.17. Auriga (Matt BenDaniel, http://starmatt.com).

The second variable star of interest is **Beta (β) Aurigae**. This is a bright star that is a good example of the Algol type of variable star, which is due to stars eclipsing each other. A spectral class A2 signifies that the hydrogen lines in the spectrum are now at their strongest. Our final variable star, **RT Aurigae**, is a classical Cepheid variable star that changes magnitude from 5.0 to 5.8 in 3.72 days. Its maximum brightness occurs over 1.5 days and its decline takes 2.5 days. Oddly enough, as it changes in brightness it also changes its spectral classification from F4 to G1.

Figure 4.18. NGC 1664 (Harald Strauss, AAS Gahberg).

No discussion of Auriga would be complete without a mention of its most famous denizen, **Capella** (**Alpha** (α) **Aurigae**). It is literally a beacon in the northern winter sky and cannot be mistaken. High in the sky in winter Capella, the sixth-brightest star in the sky, has a definite yellow color, reminiscent of the Sun's own hue. It is in fact a spectroscopic double, and is thus not split in a telescope; however, it has a fainter 10th magnitude star about 12 arcseconds to the southeast, at a PA of 137°. This is a red dwarf star, which in turn is itself a double (only visible in larger telescopes). Thus Capella is in fact a quadruple system. When seen through any optical instrument, it blazes forth and is a stunning sight.

Star clusters are plentiful in Auriga, and some are very spectacular indeed! There are also a lot of small and faint clusters that you can come across if you casually scan the area. It is always a delight to be observing and to suddenly see a cluster, or any object really, that you do not recognize and which may not even be cataloged. Our first cluster is **NGC 1664** (**Herschel 59**), which is a nice bright cluster, but loosely structured and best seen with an aperture of 20 cm (see Figure 4.18). It appears as an enrichment of the background Milky Way star field (see Star Chart 4.13). There is a 7th magnitude star within the cluster but it is not a true member, and the glare from the star can sometimes make observation of the cluster difficult. Increasing the aperture will show progressively more stars, as is to be expected.

Our next open cluster is **NGC 1778** (**Herschel 61**) (see Figure 4.19). Although this is a fairly bright cluster, it is so sparse and spread out that it will require some careful observation to be located. In small telescopes it is a northwest to southeast group of 15 stars, and even in larger telescopes it is difficult to discern from the Milky Way background (see Star Chart 4.14).

On the other hand, **NGC 1857** (**Herschel 33**) is a very rich cluster containing several small chains of stars with starless voids located within and around it (see Figure 4.20). The brightest member of the cluster is a nice orange-tinted star, but its glare can overpower the many fainter stars. Some observers try to occult the bright star so that it is obscured, thus allowing the fainter stars to be observed (see Star Chart 4.14).

A nice cluster set within a lovely rich part of the Milky Way is **NGC 1893**. It lies within a triangle-shaped group of three 8th magnitude stars and comprises about 15 stars ranging in magnitude from 9th to 11th, all within a 10 arcminutes area (see Figure 4.21). As in the case of many clusters, increasing the telescope aperture increases the number of stars seen.

I include this cluster as a test for those of you with big instruments. **NGC 1883** can easily be missed, as it is a faint patch of hazy starlight (see Star Chart 4.13). It lies some 3 arcminutes to the northeast of a 10th magnitude star and is a difficult object due to its small size and inherent faintness of magnitude 12 (see Figure 4.22).

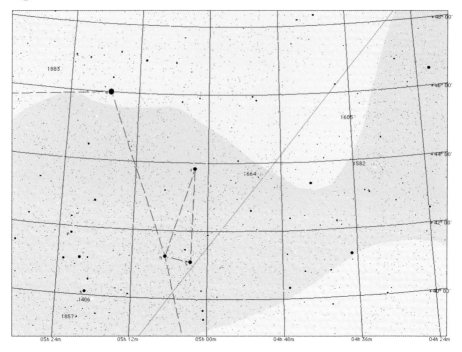

Star Chart 4.13. NGC 1664; NGC 1883.

Now for the first of our Messier objects, the open cluster **NGC 1912** (**Messier 38**) (see Star Chart 4.15). This is one of the three Messier clusters in Auriga, and is visible to the naked eye. It contains many A-type main-sequence and G-type giant stars, with a G0 giant being the brightest, of magnitude 7.9. It is elongated in shape with several double stars and voids within it. Seen as a small glow in binoculars, it is truly lovely in small telescopes (see Figure 4.23). In medium telescopes it is a rich group of about 100 stars of 9th magnitude and fainter. Some observers see a π-shaped asterism within the cluster. Do you?

Figure 4.19. NGC 1778 (Harald Strauss, AAS Gahberg).

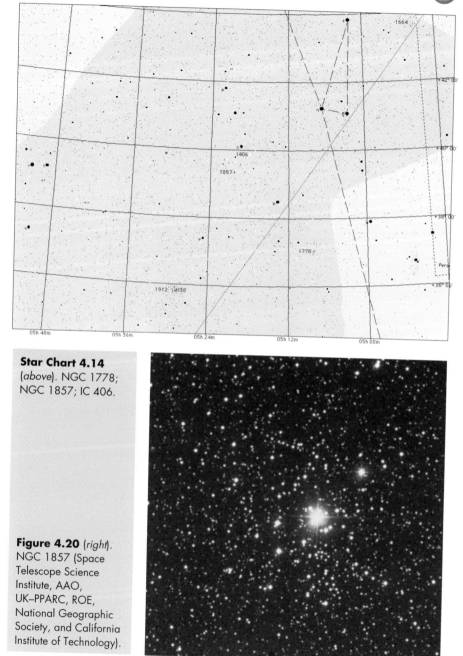

Star Chart 4.14 (*above*). NGC 1778; NGC 1857; IC 406.

Figure 4.20 (*right*). NGC 1857 (Space Telescope Science Institute, AAO, UK–PPARC, ROE, National Geographic Society, and California Institute of Technology).

Also look out for the neglected cluster **NGC 1907** that lies south of M38. Messier 38 is an old galactic cluster with a star density calculated to be about eight stars per cubic parsec. It is a compact cluster of about 30 stars ranging in magnitude from 9th to 12th.

Figure 4.21. NGC 1893 (Harald Strauss, AAS Gahberg).

Figure 4.22. NGC 1883 (Harald Strauss, AAS Gahberg).

The next Messier cluster is **NGC 1960 (Messier 36)**. At about half the size of M38, it can be seen as a faint glow in binoculars (see Star Chart 4.15). It is a large, bright cluster with about sixty 8th magnitude and fainter stars, and measurements indicate that it is ten times farther away then the Pleiades. It contains a nice double star at its center. Owing to the

Figure 4.23. Messier 38 (Harald Strauss, AAS Gahberg).

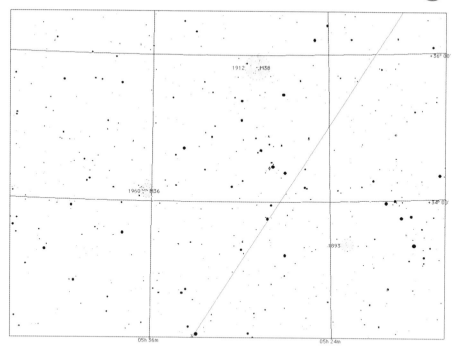

Star Chart 4.15. NGC 1893; Messier 36; Messier 38.

faintness of its outlying members it is difficult to ascertain where the cluster ends. It is visible to the naked eye when conditions allow (see Figure 4.24).

Finally there is the magnificent open cluster: **Messier 37 (NGC 2099)** (see Star Chart 4.16). How to describe it? Well, in a word – superb! It is without a doubt the finest cluster in Auriga. It really can be likened to a sprinkling of stardust, and some observers liken it to a scattering of gold dust (see Figure 4.25). It contains many A-type stars and several red

Figure 4.24. Messier 36 (Rolf Löhr , AAS Gahberg).

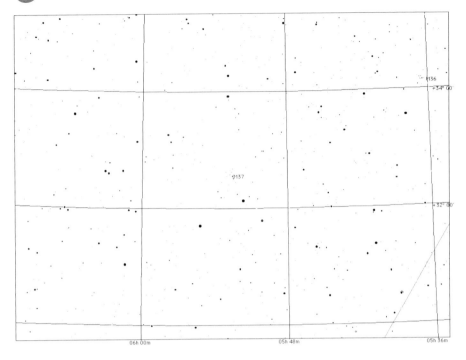

Star Chart 4.16. Messier 37.

giants and is visible in all apertures, from a soft glow with a few stars in binoculars, to a fine, star-studded field in medium-aperture telescopes. In small telescopes using a low magnification it can actually appear as a globular cluster. The central star is colored a lovely deep red, although several observers report it as a much paler red, which may indicate that it is a variable star. Out of all three Messier clusters it is the more compressed object and like its brothers is visible to the naked eye.

Figure 4.25. Messier 37 (Harald Strauss, AAS Gahberg).

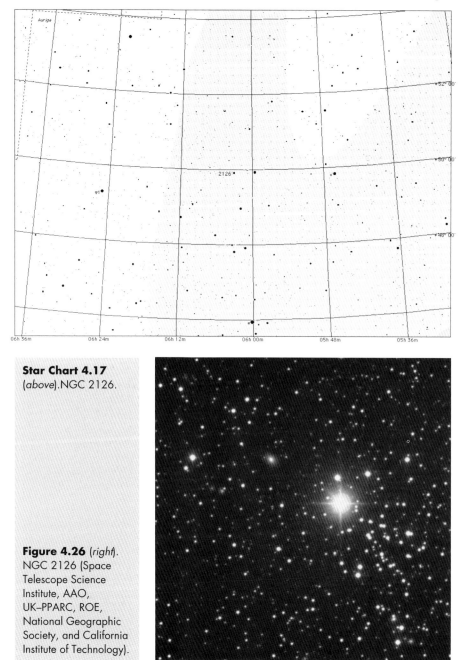

Star Chart 4.17
(*above*).NGC 2126.

Figure 4.26 (*right*).
NGC 2126 (Space
Telescope Science
Institute, AAO,
UK–PPARC, ROE,
National Geographic
Society, and California
Institute of Technology).

After the magnificence of the above clusters, now back to the more familiar, and that could include **NGC 2126** (**Herschel 68**) (see Star Chart 4.17). This cluster has been described as diamond dust on black velvet. Admittedly it is a very faint but nice cluster, although it can prove a challenge to find (see Figure 4.26).

Figure 4.27. NGC 2281 (Rolf Löhr, AAS Gahberg).

A fine asterism is **Harrington 4,** consisting of the stars **16, 17, 18, 19,** and **IQ Aurigae.** To the naked eye it will appear as a faint hazy object near the center of the constellation. In binoculars several more faint stars join the group to form a very pleasing sight. A faint cluster that is surprisingly about the same angular size as the full moon is **Collinder 62.** It has been given an integrated magnitude of 4.2, but this is very misleading as most of its stars are 8th magnitude and fainter. The culprit is a 5th -star that falsely boosts the overall cluster magnitude. However, it is still worth a brief visit, and lies about 5° south of Capella.

A cluster that is often overlooked is **Stock 10.** Although it does not have many stars in it, it does stand out rather well against the background, as several of its members are magnitude 8 or brighter. It lies some 4° north of M36 and 4° northwest of Theta Aurigae. Try to search out this hidden but nice cluster. Our final open cluster is another often passed-over object, **NGC 2281,** which, when viewed through a medium-aperture telescope, is a bright, but loose cluster comprised of around 25–30 stars ranging in magnitude from about 8 to 10 (see Figure 4.27).

There are some nebulae in Auriga, but most are faint and small. We shall of course discuss those that are visible to the majority of amateur astronomers.

The first nebula to visit is IC 410 that will need telescopes of at least 30 cm aperture and an [OIII] filter in order to be observed (see Star Chart 4.18). It lies within the cluster **NGC 1893** and is a very faint and irregularly shaped object. A nebula that can best be seen in moderate-aperture telescopes is **IC 417.** However, in order for you to observe this emission nebula the use of averted vision will be required. Then it will appear as a very faint ghostly haze with a few faint stars located within it (see Star Chart 4.18). Filters are again useful with the emission nebula, as are perfect seeing conditions.

Comprising both an emission and a reflection nebula is **NGC 1931 (Herschel 261).** Located in a nice rich star field it will appear as a small and round hazy glow, about 1 arcminute across, encompassing a triangle of faint stars (see Star Chart 4.18). In larger apertures a slight brightening can be discerned at the nebula's center (see Figure 4.28).

We conclude this section on emission and reflection nebulae with probably the most famous[6] nebula in Auriga – IC 405 (**Caldwell 31**) (see Star Chart 4.18). Also known as the **Flaming Star Nebula,** this is a very hard reflection nebula to observe. It is actually several nebulae including IC 405, 410 and 417, plus the variable star AE Aurigae. Narrow-band filters are justified with this reflection nebula, as they will highlight the various components (see Figure 4.29). A challenge to the observer!

---

[6] It is ironic that the most famous nebula is also probably the most difficult to observe. Such is the life of an astronomer.

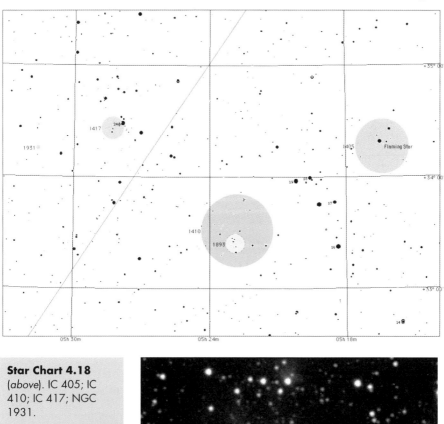

Star Chart 4.18 (*above*). IC 405; IC 410; IC 417; NGC 1931.

Figure 4.28 (*right*). NGC 1931 (Harald Strauss, AAS Gahberg).

Now for something rather unique, the supernova remnant (SNR) **Sh2–224**. Most of the SNR lies within the northern part of the constellation Taurus, but the part we are interested in can be seen best with the use of an ultra-high-contrast filter, and possible a high magnification. A problem occurs due to the presence of an 8th magnitude star, the glare from which tends to overpower the faint SNR. It is, however, a large nebula and consists of two very long sections that lie in an east–west orientation.

There are a few planetary nebulae in Auriga, but only one concerns us, and that is **IC 2149** (**PK166+10.1**). It is a bright and easily seen object that has a very marked elliptical shape, some 12 × 8 arcseconds. In small and medium-aperture telescopes it will appear stellar in character, but a high enough magnification may reveal its true nature. In larger telescopes it will show a bluish disk and the central star may be glimpsed from time to time, weather permitting of course (see Star Chart 4.19).

Figure 4.29. IC 405 (Space Telescope Science Institute, AAO, UK–PPARC, ROE, National Geographic Society, and California Institute of Technology).

Star Chart 4.19. IC 2149.

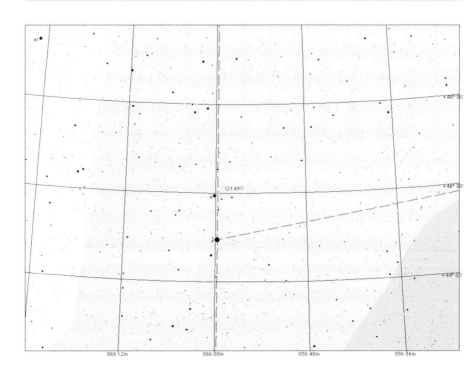

This ends our visit to the Milky Way regions of Auriga, and it is something of a first for us, because you will have noticed that we have not covered one class of celestial object – galaxies. Amazingly, there are none that are visible in telescopes that are most used by amateurs. Those of you who have very, very large instruments will be able to see a few, but the rest of us, alas, will not. Who would have thought it?

4.3 Taurus

Always a favorite constellation to look out for, **Taurus** rides high in the winter sky for northern observers, and low in the summer sky for southern observers (see Star Chart 4.20). You may be surprised to know that the Milky Way passes this way, but it does, albeit through its northern and easternmost regions. But before you get too excited, let me tell you now that unfortunately, it does not include Taurus's most famous object, Messier 45, the Pleiades. So, alas, we will not discuss them at all.[7] On the plus side, there are still some nice objects to look at and those are the ones we shall discuss below.

Star Chart 4.20. Taurus.

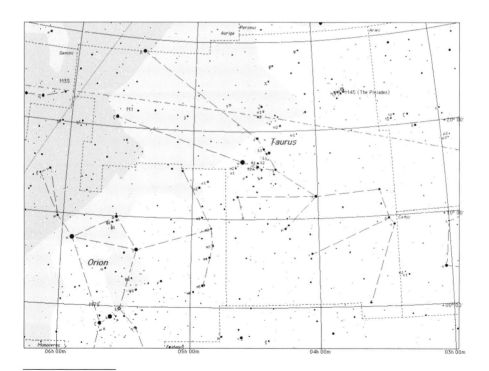

---

[7] A full description of M45 and many other celestial objects can be found in my book, *Field Guide to the Deep Sky Objects*, Springer-Verlag.

The Milky Way is faint here, but is still worth scanning with binoculars or a small telescope, as many "just-resolved" star fields can be observed. The constellation transits at the end of November.

Our first star to look at is the variable star **RW Tauri**. This is an eclipsing binary star, that lies about 1° to the northwest of **41 Tauri**. The primary star, a B8 type star, is totally eclipsed by its K-type subgiant every 2.76 days. This lasts for 9 hours while complete totality is for 84 minutes. The magnitude change is one of the biggest known for this type of star, 3.50 magnitudes as measured visually.[8]

A nice double star is **Phi** (φ) **Tauri**, which is perfect for even very small telescopes where it will appear as a pair of deep yellow and blue stars. It can even be split in binoculars. Another fine double is Σ (**Struve**) **559**. It lies in a nice faint star field, and is a pair of yellow stars of nearly equal magnitude. It is a good test for an 8 cm telescope. It came as a surprise to me to discover that the brightest star in the constellation, **Alpha** (α) **Tauri**, or **Aldebaran**, is actually a double star, but a very difficult one to separate owing to the extreme faintness of the companion. The companion star, a red dwarf star, magnitude 13.4, lies at a PA of 34° at a distance of 121.7 arcseconds. Aldebaran, which is the fourteenth-brightest star, is apparently located in the star cluster the Hyades. However, it is not physically in the cluster, lying as it does twice as close as the cluster members. This pale orange star is around 120 times more luminous than the Sun.

Another pleasing double star system of pale yellow stars is Σ (**Struve**) **572**. It too lies in a star-sprinkled field. Our final double star, **118 Tauri**, lies within what appears to be a very large dark region of the Milky Way, presumably an immense dust cloud. This is a pair of yellow-tinted stars that can be used as a test for small-aperture telescopes. Apparently the pair is a physical system and has a similar proper motion through space.

Now let's look at a few nonstellar objects. Our first target is the constellation's only easily observed planetary nebula, **NGC 1514 (Pk165–15.1)** (see Star Chart 4.21). Providing you use a medium to high magnification, this will be visible in, say, a telescope of 20 cm aperture where it will appear as a bright disk, diameter 1.5 arcminutes. What makes it special is that the central part is of magnitude 9.5, so it can be seen easily (see Figure 4.30).

A huge star cluster that everyone knows is **Collinder 50 (Melotte 25)**. We all know it as the **Hyades**. The Milky Way passes just through its northern reaches, and so I have decided to include it here, even though only about one-third of it actually lies upon the Milky Way.[9] It is the nearest cluster after the **Ursa Major Moving Stream,** lying at a distance of 151 light years, with an age of about 625 million years, and it is best seen with binoculars owing to the large extent of the cluster – over 5°. Hundreds of stars are visible, including the fine orange giant stars γ, δ, ε and θ¹ **Tauri**. Aldebaran is not a true member of the cluster, but is a foreground star only 70 light years away. It is visible even from light-polluted urban areas – something of a rarity! Even though the cluster is widely dispersed both in space and over the sky, it nevertheless is gravitationally bound, with the more massive stars lying at the center of the cluster. Another nice cluster is **NGC 1746 (Melotte 28)** (see Star Chart 4.22). This is another large and scattered cluster, visible on clear nights with the naked eye (see Figure 4.31). There are over 70 stars in the cluster spread over nearly 40 arcminutes, which makes it larger than the full moon!

---

[8] The photographic magnitude change has been measured to be some 4.49 magnitudes.

[9] It is always a problem knowing what to include in the book, and what to leave out, but as I mentioned in the first chapter, I try to stick rigidly to the premise that an object can be included if it lies within the boundaries of the Milky Way as defined by the Dutch Astronomer Antonie Pannakoek (as used in the star atlas *Sky Atlas 2000.0*), who measured the approximate brightness levels of the Milky Way. Anything outside of this is not mentioned. This does leave out a lot of famous and bright objects, but if I were to include them, this book would run to several volumes, and possibly a few bank overdrafts.

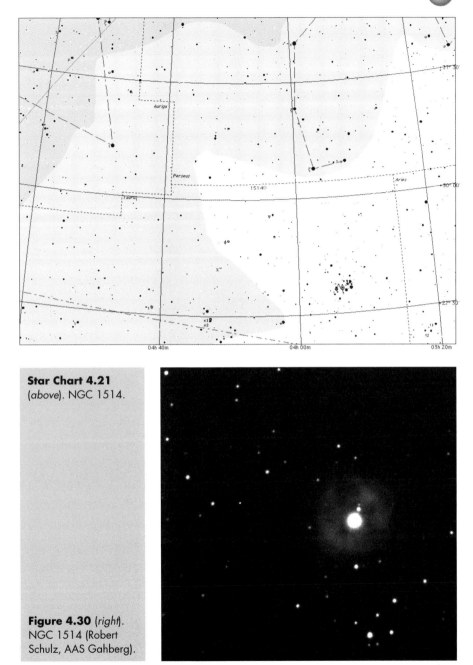

Star Chart 4.21 (*above*). NGC 1514.

Figure 4.30 (*right*). NGC 1514 (Robert Schulz, AAS Gahberg).

Within the cluster are two other smaller ones, each with its own classification – **NGC 1750** and **NGC 1758**. This makes it difficult to determine accurately the true diameter of the cluster, not that it matters to us, as it is still nice to observe. Two nice clusters that lie on the border with Orion are **NGC 1807** and **NGC 1817** (see Star Chart 4.23). Both can be seen

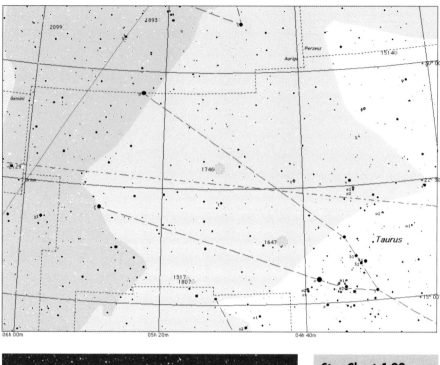

Star Chart 4.22
(*above*). NGC 1746.

Figure 4.31 (*left*).
NGC 1746 (Space
Telescope Science
Institute, AAO,
UK–PPARC, ROE,
National Geographic
Society, and California
Institute of Technology).

in telescopes of 20 cm, where the former will appear as a loose collection of about thirty 9th magnitude and fainter stars lying within a 12 arcminute area (see Figure 4.32). The latter lies within the same field as NGC 1807 and comprises a chain of about 70 near-10th magnitude stars (see Figure 4.33). Although both can be glimpsed in binoculars, NGC 1807

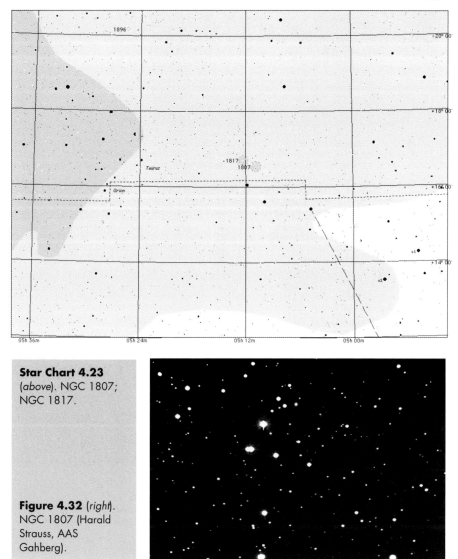

Star Chart 4.23 (*above*). NGC 1807; NGC 1817.

Figure 4.32 (*right*). NGC 1807 (Harald Strauss, AAS Gahberg).

will only show 20 stars, whereas NGC 1817 is far less obvious and may only appear as a faint glow.

Lying within a dark region of the Milky Way is the grand-sounding cluster **Dolidze–Dzimselejsvili 3**. It is fairly bright but well spread out and in truth is not really impressive, but because it is in such a star-sparse area, presumably due to the aforementioned dark region, it stands out in comparison rather well. Another of these clusters with an unfamiliar classification is **Dolidze–Dzimselejsvili 4**. Oddly, both the Herschels and Messier missed it, because it is a bright open cluster that lies about 4° to the northeast of M1 (see below). It has about 30 stars, ranging in magnitude from 6th to 10th. In binoculars you can see about a dozen of these.

Figure 4.33. NGC 1817 (Harald Strauss, AAS Gahberg).

Possibly the most famous deep-sky object in Taurus, and easily the most famous of its class, is the supernova remnant, **Messier 1**, or the **Crab Nebula (NGC 1952)** (see Star Chart 4.24). This is without a doubt the most famous SNR in the sky, and it can even be glimpsed in binoculars as an oval light of plain appearance (see Figure 4.34). With telescopes of aperture 20 cm it becomes a ghostly patch of grey light and larger-aperture telescopes will show some faint mottled structure. In all apertures (except very large – 40 cm) it will remain uniform in appearance.

Star Chart 4.24. Messier 1.

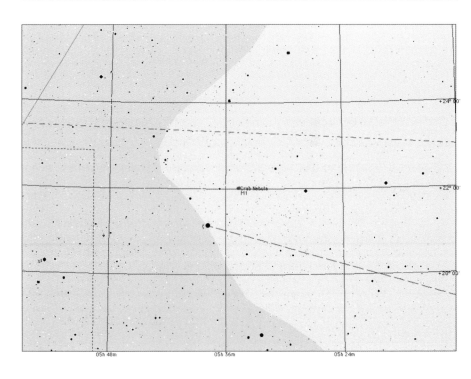

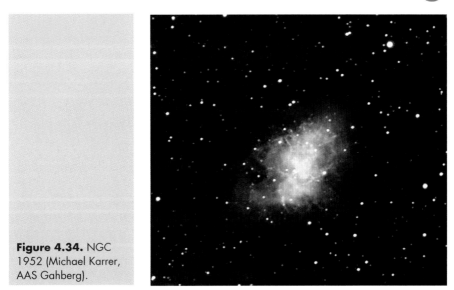

Figure 4.34. NGC 1952 (Michael Karrer, AAS Gahberg).

In 1968, in the center of the Crab nebula was discovered the **Crab Pulsar,** the source of the energy responsible for the pearly glow observed, a rapidly rotating neutron star which has also been optically detected. The Crab Nebula is a type of supernova remnant called a plerion, which, however, is far from common among supernova remnants. Another peculiar thing about SNRs is that although a lot of them have been detected, as well as quite a number of pulsars, only a meager handful of SNRs have a pulsar that is physically associated and that is presumably, the precursor star.

There is another SNR in Taurus, Sh2–22, an object I mentioned earlier in the section on Auriga. It is, however, not readily visible, and to my knowledge has not been seen visually with medium-aperture telescopes. It would, I believe, make an excellent CCD-imaging target. And with that we leave Taurus, and move on to Gemini.

4.4 Gemini

Our penultimate constellation on our journey through the Milky Way is **Gemini,** a famous and distinct constellation for northern observers (see Star Chart 4.25). The Milky Way passes though most of it, and only its easternmost section is left out. Alas, that means that its two famous brothers, Castor and Pollux, are omitted, but there are plenty of other fine deep-sky objects to interest us. The center of the constellation actually transits at the very beginning of January, but the center of the areas we are most concerned with transit at the end of December, thus its inclusion here. It is not too low for southern observers and so it is easy to make observations. Without further ado, let's begin.

There are a few interesting variable stars we should look at: **BU Geminorum** and **U Geminorum.** The first is an irregular variable star that is perfect for observation with binoculars or a small telescope, as this red star varies in magnitude from 5.7 to 7.5. What is intriguing is that the period of variability is not regular so you can never tell when it will alter. The latter star is the prototype of what are called cataclysmic variable stars. Normally the star is around 15th magnitude and so is very faint for most of us, but, every 100 days or

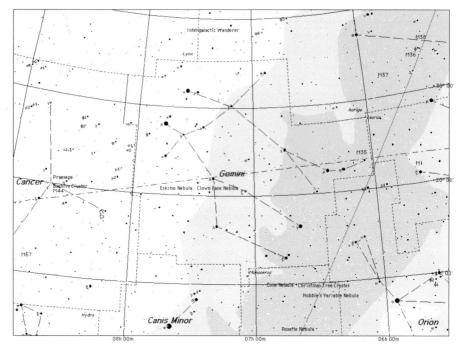

Star Chart 4.25. Gemini.

thereabouts its magnitude drastically brightens to between 8 and 9 in as little as one day. This is quite some change, so should be watched for. An interesting star is **Zeta (ζ) Geminorum**, one of the brightest examples of a Cepheid variable star in the entire sky. Its magnitude changes from 3.6 to 4.1 in about 10.15 days and so it is perfect for binocular and even naked-eye observation. Stars that can be used as comparisons and that lie close by are **Kappa (κ) Geminorum** and **Upsilon (υ) Geminorum**, which are at magnitudes 3.57 and 4.06 respectively.

As to be expected, there are many fine double stars in this region of the Milky Way. There are several that can be easily resolved in small telescopes and of these the following are some of the best representatives. Our first is the nice color-contrasted pair, **15 Geminorum**. This consists of a pale yellow and a blue star easily split in a 10 cm telescope. Our next double star is **20 Geminorum**, which will need a telescope of 10 cm in order to be fully appreciated, where it will appear as two pale yellow and white stars of nearly equal brightness. Another easily split double system is **38 Geminorum**, which has a lemon-tinted star paired with a pale blue star. Our final star in this short list is **Σ (Struve) 1108**, which couples a nice yellow star with a seemingly tiny blue-tinted companion.

Our final double star will need a slightly larger aperture in order to be truly appreciated: **Delta (δ) Geminorum**. This consists of a nice bright yellow star with a pale blue secondary. What makes this system so special is reports that in large telescopes, say 30 cm or more, a very distinct color contrast becomes apparent between the closer pair of Delta (δ) Geminorum, which reportedly has the colors of yellow and rare reddish-purple! I have to admit that I could never see this contrast.

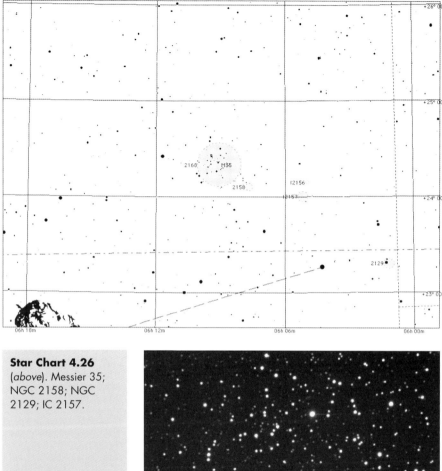

Star Chart 4.26 (*above*). Messier 35; NGC 2158; NGC 2129; IC 2157.

Figure 4.35 (*right*). Messier 35 (Harald Strauss, AAS Gahberg).

As the Milky Way passes through here there are, as we have come to expect, several fine open clusters. Many are visible in small or medium-aperture telescopes, and some need larger apertures, but we shall not concern ourselves with these. Instead we shall look at the ones that are easy to observe, and this includes one that many believe is one of the finest in the winter sky, **Messier 35 (NGC 2168)**. This is without a doubt one of the most magnificent clusters in the sky (see Star Chart 4.26). Messier 35 is visible to the naked eye on clear winter nights with a diameter as big as that of the full moon, when it seems as if the cluster is just beyond being resolved (see Figure 4.35). Many more stars are visible in binoculars set against the hazy glow of unresolved members of the cluster. With telescopes, the magnificence of the cluster becomes apparent, with many curving chains of stars. This is a cluster that should be on every observer's list of things to look at.

Figure 4.36. NGC 2158 (Harald Strauss, AAS Gahberg).

Located in a lovely star field just southwest of M35 is the cluster **NGC 2158 (Collinder 81)** (see Star Chart 4.26). Lying at a distance of 160,00 light years, this is one of the most distant clusters visible using small telescopes, and lies at the edge of the Galaxy. It needs a 20 cm telescope to be resolved, and even then only a few stars will be visible against a background glow (see Figure 4.36). It is a very tight, compact grouping of stars, and something of an astronomical problem. Some astronomers class it as intermediate between an open cluster and a globular cluster, and it is believed to be about 800 million years old, making it very old as open clusters go.

A cluster that is visible in binoculars is **NGC 2129 (Collinder 77)**, although it will just look like a hazy patch (see Star Chart 4.26). In medium-aperture telescopes using a medium magnification, it will have a bright appearance with stars ranging from 10th to 12 magnitude (see Figure 4.37). It is dominated by two stars of magnitudes 7.4 and 8.6 which probably give rise to its rather high and admittedly biased integrated magnitude.

On the other hand, the clusters **IC 2157 (Collinder 80)** and **NGC 2304 (Herschel 2)** are rather faint and small, even in telescopes of 20 cm aperture. The former has only a small number of bright stars and quite a few faint ones, whereas the latter will appear as a faint haze of about 20 stars (see Star Chart 4.26). Nevertheless they are objects to be sought out.

A rich and very nice cluster is **NGC 2266 (Herschel 21)**, which has a very distinct triangular shape, consisting of about 50 stars (see Star Chart 4.27). In larger telescopes it becomes even better and is well worth a visit (see Figure 4.38). A cluster that lies within a

Figure 4.37. NGC 2129 (Harald Strauss, AAS Gahberg).

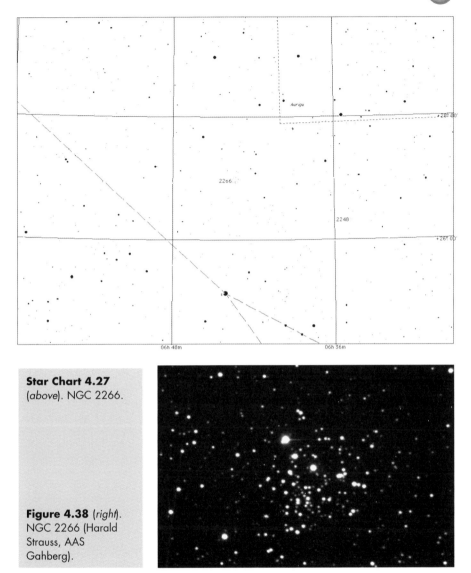

Star Chart 4.27 (*above*). NGC 2266.

Figure 4.38 (*right*). NGC 2266 (Harald Strauss, AAS Gahberg).

rather rich region of the Milky Way is **NGC 2331 (Collinder 126)**, which is good news for us as observers, as visually the cluster is not that impressive. It has about 25 faint stars with two-thirds of them at magnitude 10 and 11.

A difficult cluster to locate and nearly impossible in small binoculars is **Collinder 89**.[10] There are about 10–20 stars in the cluster, half a dozen of which can be observed in small telescopes and binoculars. To locate it can be a problem but it is between 9 Geminorum and 10 Geminorum, so that may help you.

Our penultimate open cluster is **NGC 2355 (Herschel 6)**, which once again for clusters in this region is faint and has about 30 members loosely scattered around a 9 arcminutes

---

[10] But not completely so. I eventually found it on a clear transparent night. Try it for yourselves.

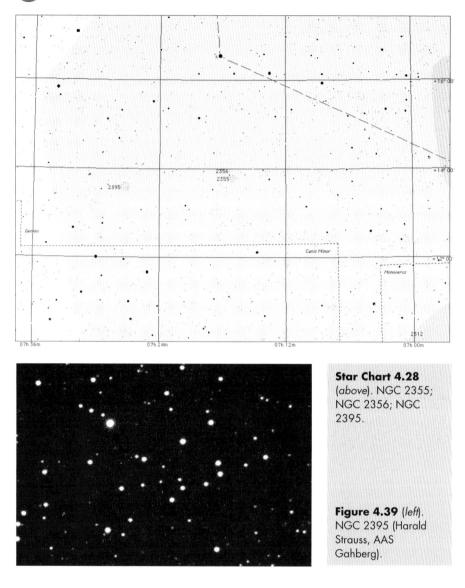

Star Chart 4.28 (*above*). NGC 2355; NGC 2356; NGC 2395.

Figure 4.39 (*left*). NGC 2395 (Harald Strauss, AAS Gahberg).

region. It will need a telescope of 20 cm or more to be appreciated. Finally there is **NGC 2395** (**Collinder 144**), which makes a nice change, as it is fairly bright and consists of around 40 stars of magnitude 9 and fainter. What makes this cluster so nice is that it is apparently split into two sections (see Figure 4.39). A northern group is large and somewhat more concentrated than its southern counterpart. In larger telescopes the two groups seem linked by a chain of stars (see Star Chart 4.28).

There are several nebulae both planetary and emission, and galaxies that are located here, but all are faint and so difficult for us to observe. All that is except for one object, the planetary nebula known as the **Eskimo Nebula**. It is also known as the **Clown Face Nebula**, **NGC 2392** (**Caldwell 39**) (see Star Chart 4.29). This is a small but famous planetary nebula that can be seen as a pale blue dot in a telescope of 10 cm although it can be glimpsed in

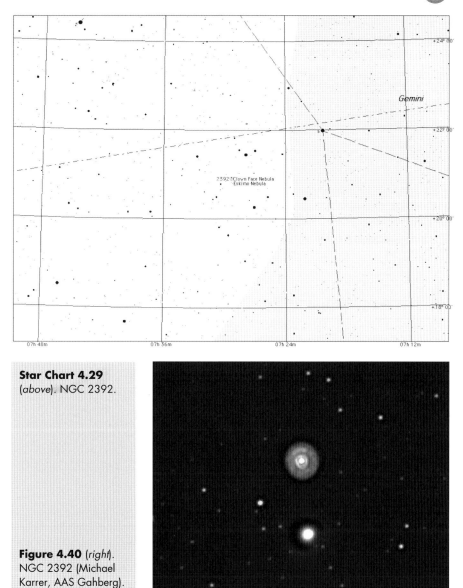

Star Chart 4.29 (*above*). NGC 2392.

Figure 4.40 (*right*). NGC 2392 (Michael Karrer, AAS Gahberg).

binoculars as the apparent southern half of a double star (see Figure 4.40). Higher magnification will resolve the central star and the beginnings of its characteristic "Eskimo" face. With aperture of 20 cm, the blue disk becomes apparent. The shell, ring and halo structure will need apertures of 40 cm in order to become easily resolvable. Research indicates that we are seeing the planetary nebula pole-on, although this is by no means certain. Its distance is also in doubt, with values ranging from 1600 to 7500 light years.

And with that we must, alas, say goodbye to Gemini, and move on to our final constellation, Orion.

4.5 Orion

We now come to the last part of our journey through the winter Milky Way, but take heart, as the constellation we now look at is regarded as one of the finest in the entire sky. **Orion** looms large in the winter sky and can be seen equally well by both northern and southern observers, as it straddles the celestial equator (see Star Chart 4.30). The constellation transits in the middle of December and so is a wonderful place to scan on cold and frosty winter nights.

The Milky Way runs through the easternmost regions of the constellation, and a look at an star atlas will show fingers of the Milky Way reaching into the constellation. The true meaning behind this is best seen on deep images of the region that show most of the constellation swathed in dust and gas. In fact, nearly all of Orion is a vast stellar nursery, as can be witnessed by the plethora of hot blue-white stars recently born from the nursery that is the Orion Complex. It is also a constellation of superlatives: the sky's finest emission nebula, and most distinctly shaped dark nebulae. But I am getting ahead of myself, so let's begin our final journey.

There is a multitude of double and multiple stars here and I shall just mention what I regard as the most impressive.

Our fist is **Rho (ρ) Orionis**, a lovely pair of yellow-orange and white stars that are easily split in the smallest of telescopes. Our next is also an easy system for all apertures: **Delta (δ) Orionis**. This comprises a bluish-white pair of stars. The brighter star is also an eclipsing binary that changes by 0.2 of a magnitude in 5.73 days. While this may not be easy to

Star Chart 4.30. Orion.

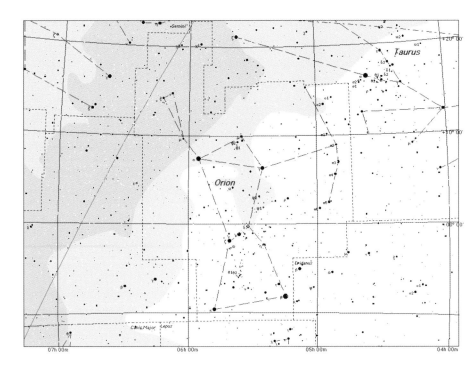

see visually, it is well within the range of amateur photometric equipment. As this star is also very close to the celestial equator, it is perfect for determining the field of view of an eyepiece. To do this is easy. Set up the telescope so that the star will pass through the center of the field of view. Now, using any fine controls you may have, adjust the telescope in order to position the star at the extreme edge of the field of view. Turn off any motor drives and measure the time t it takes for the star to drift across the field of view. This should take several minutes and seconds, depending on the eyepiece used. Then multiply this time by 15, to determine the apparent field diameter of the eyepiece, which is conveniently also in minutes and seconds – but minutes and seconds of arc, rather than of time.[11]

What Orion has in plenty are glorious multiple stars and our first of these is **Sigma (σ) Orionis**. This is a multiple star system of one white and three bluish stars. The two brightest stars are actually field stars and are not physically associated with the system. Then there is **Zeta (ζ) Orionis**. This is a nice triple system of blue and white and very pale red stars. Note that it is located among and near several bright and dark nebulae, and is a pointer to the famous Horsehead Nebula, which we will mention later. Another fine multiple system is **Iota (ι) Orionis** which is a nice color-contrasted triple system and also a test for small telescopes. There is also some nebulosity in and around the star that is discussed below. The stars are colored white with delicately tinted blue and red companions. Two more delightful multiples are **Eta (η) Orionis** and **Lambda (λ) Orionis**. The former is also known as **Dawes 5**, and is a wonderful system. Even under high magnification, the two brighter members will appear as two white disks in contact. The star is also a spectroscopic binary. The latter is a very colorful object with many colors being attributed to it. Basically it is a nice quadruple star system of white and blue stars. However, various observers have reported them as yellowish and purple and pale white and violet. What colors do you see?

The most spectacular multiple star in Orion and possibly one of the most famous in the entire sky is of course the **Trapezium, Theta¹ (θ¹) Orionis**. The four stars which make up the famous quadrilateral are set among the wispy embrace of the Orion Nebula, M42, one of the most magnificent sights in any telescope (see Star Chart 4.31). They are very young stars recently formed from the material in the nebula, and so should all appear bright white, but the nebula itself probably affects the light that is observed, so the stars appear as off-white, delicately tinted yellowish and bluish. Other observers have reported the colors as pale white, faint lilac, garnet and reddish! It is believed that these four stars contribute nearly all the radiation that makes the Orion Nebula glow. It is well worth spending an entire evening just observing the region. Incidentally this star is always a good test for small telescopes. A glorious sight and in my opinion one of the highlights of the night sky.

Before we leave the individual stars and move on to deep-sky objects, there is one star that must be mentioned, as it is a wonder to behold: **Betelgeuse**, or **Alpha (α) Orionis**. This is the tenth-brightest star in the sky, and always a favorite among observers. The distinctly orange-red star is a giant variable with an irregular period, and recent observations by the *Hubble Space Telescope* have shown that it has features on its surface that are similar to sunspots, but much larger, covering perhaps a tenth of the surface. It also has a companion star, which may be responsible for the nonspherical shape it exhibits. Although a giant star, it has a very low density and a mass only 20 times greater than the Sun's, which together mean that the density is in fact about 0.000 000 005 that of the Sun. A lovely sight in a telescope of any aperture; subtle color changes have been reported as the star goes through its variability cycle.

---

[11] If you have to use a star which does not lie on or close to the celestial equator, then the formula $15t \cos \delta$, where δ is the declination of the star, can be used to find the apparent field diameter of the eyepiece in minutes and seconds of arc.

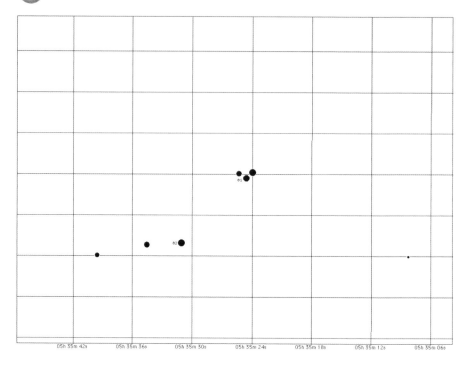

Star Chart 4.31. Trapezium.

It may be wise at this point to mention that most of the stars in the constellation down to about magnitude 3.5 except for **Gamma (γ) Orionis** and **Pi³ (π³) Orionis** are part of the Orion Association. The nebula M42, along with many dark, emission and reflection nebulae, are all located within a vast **Giant Molecular Cloud**, which is the birthplace of all the O-type and B-type supergiant, giant, and main-sequence stars in Orion. The association is believed to be about 800 light years across and 1000 light years deep. When you look at this association you are in fact looking deep into our own spiral arm, which incidentally, is called the **Cygnus–Carina Arm**.

Now let's look at some open clusters. Our first is the large and bright cluster, **Collinder 65**, which can be found at the northern part of Orion. In fact several of the cluster's members lie in Taurus. It is large, at nearly 2° and so is best suited to binocular observation. The cluster tends to form a diamond-shaped group of stars which range from 6th to 8th magnitude.

Lying on the border of an enormous patch of dark obscuration is the cluster **Collinder 70**. We have all seen it, but few realize it is a cluster. It comprises the three 2nd magnitude stars that make up the belt of Orion and a collection of 5–8th magnitude stars, and quite a few fainter ones too, making the total about 100 stars. The whole group is given the cataloged name and in binoculars is a spectacular sight.

Another cluster that is perfect for binoculars is **Collinder 69**. This cluster surrounds the 3rd magnitude stars λ Orionis, and includes φ⁻¹ and φ⁻² Orionis, both 4th magnitude stars. Encircling the cluster is the very faint emission nebula **Sharpless 2–264**, only visible using averted vision and an [OIII] filter with extremely dark skies.

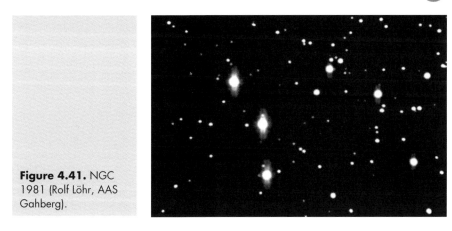

Figure 4.41. NGC 1981 (Rolf Löhr, AAS Gahberg).

A nice, bright, coarse cluster, lying about 1° north of Messier 42 is **NGC 1981** (**Collinder 73**). Consisting of around eight or nine stars it can be seen in binoculars, while the remaining stars are a hazy background glow. In moderate telescopes, the most striking feature is two parallel rows of stars (see Figure 4.41).

A small but delightful cluster is **NGC 2169** (**Collinder 38**) (see Star Chart 4.32). This is a small but bright cluster and some observers find it hard to believe that this scattering of stars has been classified as a cluster at all! Easily visible in binoculars, the stars appear to

Star Chart 4.32. NGC 2169.

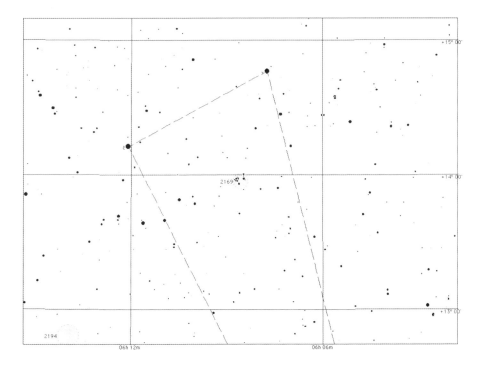

Figure 4.42. NGC 2169 (Harald Strauss, AAS Gahberg).

range in magnitude from about 8 to 10. Also, binoculars will show the four brightest members to be surrounded by faint nebulosity – sometimes! You will need very clear nights and very clean optics to detect this faint nebula (see Figure 4.42).

We are now going to look at some objects in Orion that are truly spectacular: the emission, reflection and dark nebulae. Some of them will require no more than binoculars, while others need a large telescope and probably an [OIII] filter. But whatever equipment you have you will not be disappointed. In fact, one object that is so amazing, the Orion Nebula, can actually be seen with the naked eye (see Figure 4.43).

So let's start by looking at the most impressive emission nebula in the heavens, the **Orion Nebula, Messier 42 (NGC 1976)** (see Star Chart 4.33). It is visible to the naked eye as a barely resolved patch of light and shows detail from the smallest aperture upwards. It is really one of those objects where words cannot describe the view seen. In binoculars its pearly glow will show structure and detail, and in telescopes of aperture 10 cm the whole field will be filled. The entire nebulosity is glowing owing to the light (and thus energy) provided by the famous Trapezium stars located within it (see Figure 4.44).

What is also readily seen along with the glowing nebula are the dark, apparently empty and starless regions. These are still part of the huge complex of dust and gas, but are not glowing by the process of fluorescence – instead they are vast clouds of obscuring dust. The emission nebula is one of the few that shows definite color and many observers report seeing a greenish glow, along with pale grey and blue. Many amateurs state that with very large apertures of 35 cm a pinkish glow can be seen. Located within the nebula are the famous **Kleinmann–Low Sources** and the **Becklin–Neugebauer Object**, which are believed to be dust-enshrouded young stars. The whole nebula complex is a vast stellar nursery. M42 is at a distance of 1700 light years, and is about 40 light years in diameter. Try to spend a long time observing this object – you will benefit from it. Many observers just let the nebula drift into the field of view. This is the sort of object that makes part-time astronomers turn into full-time astronomy devotees!

Some other nebulae that actually suffer from being so close to M42 are the emission and reflections nebulae, **NGC 1973, NGC 1975** and **NGC 1977**. They lie between M42 on their south and the cluster NGC 1981 to their north (see Star Chart 4.33). However, they are also difficult to see because the glare from the star tends to make observation difficult. Amateurs who own large telescopes will notice that the area is immersed in swathes of dark and light nebulosity (see Figures 4.45 and 4.46).

Another nebula that also has a very fine cluster within it is **NGC 1980**, located to the south of M42. Set in a lovely part of the Milky Way the cluster comprises about 30 stars, within which is the triple star mentioned above, Iota (ι) Orionis. The nebulosity that sur-

Figure 4.43. Nebulae in Orion (Matt BenDaniel, http://starmatt.com).

rounds the star is about 5 arcminutes across and there is a further area about the same size some 8 arcminutes to the southwest.

Visible in binoculars is the emission nebula **Messier 43** (**NGC 1982**), which is relatively large and surrounds a 7.5 magnitude star (see Star Chart 4.33). The emission nebula is part of the M42 complex, and some observers find it difficult to discriminate between them. Visible to the north of M42, it takes magnification well, and will show many intricate details (see Figure 4.47).

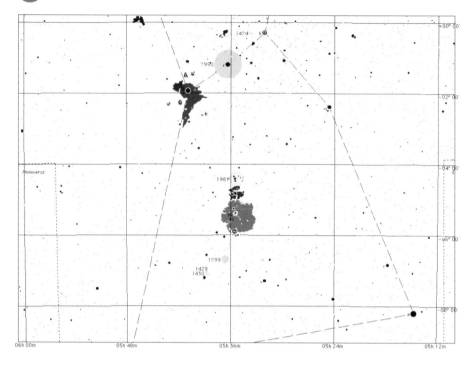

Star Chart 4.33. NGC 2023; NGC 1981; NGC 1428; NGC 1990; NGC 1999; Barnard 33; NGC 1973-75-77; NGC 2024; Messier 42; Messier 43; Messier 78.

All of the objects mentioned above can fit into one eyepiece at low power and the resulting sight is often mesmerizing.

But we haven't finished yet, as more nebulae can also be seen. In telescopes of aperture 20 cm, the small but bright emission nebula **NGC 1999** can be seen, resembling a planetary nebula. It even has a star in its central region of magnitude 9.4 (see Figure 4.48). It lies about 1° south of M42 (see Star Chart 4.33).

One nebula that suffers from being close to a bright star is **NGC 2024**. This difficult nebula lies next to the famous star Zeta (ζ) Orionis, which is unfortunate as the glare from the star makes observation difficult (see Star Chart 4.33). It can, however, be glimpsed in binoculars as an unevenly shaped hazy and faint patch to the east of the star, providing the star is placed out of the field of view. With large telescopes and filters the emission nebula is a striking object, and has a shape reminiscent of a maple leaf (see Figure 4.49).

A bright but small emission nebula that can be seen in binoculars is **Messier 78** (NGC 2068) (see Star Chart 4.33). It has a distinct fan shape, whereas some observers liken it to a comet (see Figure 4.50). There are two 10th magnitude stars located within the nebula which can give the false impression of two cometary nuclei.

With a large-aperture telescope and high magnification, some very faint detail can be glimpsed along the eastern edge of the nebulosity, but excellent seeing will be needed in order to observe this.

Another emission and reflection nebula is **NGC 2023** that surrounds an 8th magnitude star (see Figure 4.51). It can be seen in a telescope of 30 cm and larger, but is a challenge for any smaller aperture (see Star Chart 4.33).

Figure 4.44. Messier 42 (Michael Karrer, AAS Gahberg).

There are several other reflection nebulae in Orion, but all are faint and they are not easy targets. Suffice to say that dark skies are a must in order for them to be seen. They include amongst their number **IC 430**, **IC 431** and **IC 432**. The former is about 5 arcminutes northwest of 49 Orionis, and thus the glare from the star tends to make life difficult to the observer. The remaining two nebulae can be seen as a faint bluish haze surrounding stars.

Figure 4.45. NGC 1973 (Harald Strauss, AAS Gahberg).

Figure 4.46 (*above*). NGC 1977 (Klaus Eder and Georg Emrich, AAS Gahberg).
Figure 4.47 (*below*). Messier 43 (Harald Strauss, AAS Gahberg).

Figure 4.48. NGC 1999 (Harald Strauss, AAS Gahberg).

We have discussed dark nebulae in other chapters, but none have as much effect on the observer as **Barnard 33**, the **Horsehead Nebula**. Often photographed, but very rarely observed, this famous dark nebula is very difficult to see (see Star Chart 4.33). It is a small dark nebula which is seen in silhouette against the dim glow of the emission nebula **IC 434**. Both are very faint and will need perfect seeing conditions (see Figure 4.52). This is what the British astronomer Don Tinkler has to say about it: "Dark nebulas were always a puzzle to me as I first embarked on astronomy. Until I seriously began to study astronomy most of the night sky was a mystery. What are these magnificent objects in the sky? And it

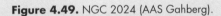

Figure 4.49. NGC 2024 (AAS Gahberg).

Figure 4.50. Messier 78 (Klaus Eder and Georg Emrich, AAS Gahberg).

was the beauty and mystery of such things that drew me into finding out more about them. One of my favorites was, and still is, the Horsehead Nebula. How can such shapes exist in space? Even though I have studied such objects, the mystery of the Horsehead Nebula still accompanies me every time I observe it". At the high resolution of an image the Horsehead appears very chaotic, with many wisps and filaments and diffuse dust. Such structures are only temporary, as they are being constantly eroded by the expanding region of ionized gas and are destroyed on timescales of typically a few thousand years. The Horsehead as we see it today will therefore not last forever and minute changes will become observable as the time passes. Such is the elusiveness of this object that even telescopes of 40 cm are

Figure 4.51. NGC 2023 (Harald Strauss, AAS Gahberg).

Figure 4.52. IC 434 (Klaus Eder and Georg Emrich, AAS Gahberg).

not guaranteed a view. The use of dark adaptation and averted vision, along with the judicious use of filters, may result in its detection. Nevertheless, have a go!

The sole planetary nebula that we can observe in Orion is **NGC 2022 (PK 196–10.1)** (see Star Chart 4.34). This is a very small, faint, grey planetary nebula, but it can be glimpsed in telescopes of 20 cm. Using larger apertures will resolve the disk appearance along with a pale greenish tint (see Figure 4.53).

Our final object in Orion, and indeed in the book, is a far from familiar one. It is an object that is often mentioned in the texts, but rarely observed: the truly vast **Sharpless 2–276**, also known as **Barnard's Loop**. This is a huge arcing loop of gas located to the east of the constellation Orion, and research suggests that it may well be the remains of a very old supernova (see Figure 4.54). It encloses both the sword and belt of Orion, and if it were a complete circle it would be about 10° in diameter. The eastern part of the loop is well defined, but the western part is exceedingly difficult to locate, and has never to my knowledge been seen visually, only being observed by the use of photography or using a CCD. Impossible to see through a telescope, recent rumors have emerged that it has been glimpsed by a select few by using either an [OIII] filter or an ultra-high-contrast filter. Needless to say, perfect conditions and very dark skies will greatly heighten the chances of it being seen. Personally I think that this object is possibly the greatest observing challenge to the naked-eye observer that we have today, especially in the light-polluted world we live in. It may be that with the ever-encroaching scourge that is light pollution, we may never see it visually again. A sobering thought.

The following constellations are also visible during these months at different times throughout the night. Remember that they may be low down and so diminished by the effects of the atmosphere. Also, you may have to observe them either earlier than midnight, or some considerable time after midnight, in order to view them.

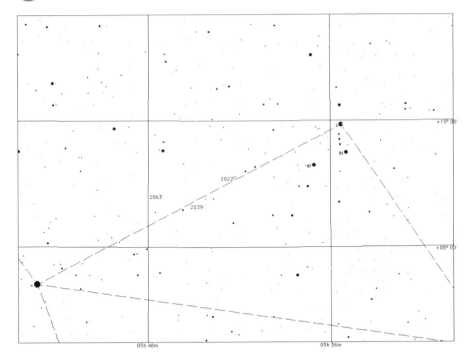

Star Chart 4.34. NGC 2023.

Northern Hemisphere

Antila, Aquila, Camelopardalis, Canis Major, Canis Minor, Cassiopeia, Cepheus, Cygnus, Delphinus, Lacerta, Lepus, Lyra, Monoceros, Perseus, Puppis, Pyxis, Sagitta, Vela.

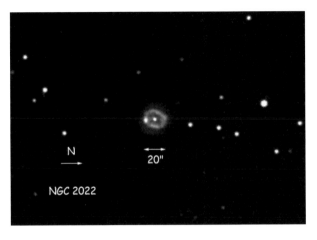

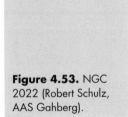

Figure 4.53. NGC 2022 (Robert Schulz, AAS Gahberg).

Figure 4.54. Barnard's Loop (Chuck Vaughn).

Southern Hemisphere

Antila, Apus, Ara, Auriga, Canis Major, Canis Minor, Carina, Centaurus, Columba, Crux, Gemini, Hydra, Lepus, Monoceros, Musca, Norma, Orion, Perseus, Puppis, Pyxis, Triangulum Australe, Vela.

- - - - - - - -

So we come to the end of the first part of our year-long voyage around our home Galaxy, the Milky Way. The southern constellations, which can be observed in the winter and spring months from northern latitudes, are described in the second book: *Astronomy of the Milky Way: The Observer's Guide to the Southern Milky Way.*

Objects in Perseus

Stars

Designation	Alternate name	Vis. mag	RA	Dec.	Description
Σ 336	Struve 336	6.9, 8.4	$03^h01.5^m$	+32° 25'	PA 8°, Sep. 8.4"
Theta (θ) Persei		4.10, 9.9	$02^h44.2^m$	+49° 13'	PA 305°, Sep. 20"
Σ 268	Struve 268	6.8, 8.1	$02^h30.8^m$	+55° 33'	PA 129°, Sep. 2.7"
Σ 270.	Struve 270	7.4, 9.2	$02^h29.4^m$	+55° 32'	PA 303°, Sep. 21.2"
Eta (η) Persei.		3.8, 8.5	$02^h50.7^m$	+55° 54'	PA 300°, Sep. 28.3"
Σ? 369	Struve 369	6.7, 8.0	$03^h17.2^m$	+40° 29'	PA 128°, Sep. 3.5"
Σ 392	Struve 392	7.4, 9.6	$03^h30.3^m$	+52° 54'	PA 347°, Sep. 25.8"
Zeta (ζ) Persei		2.9, 9.5	$03^h54.1^m$	+31° 53'	PA 208°, Sep. 12.9"
56 Persei		5.9, 8.7	$04^h24.6^m$	+33° 58'	PA 22°, Sep. 4.2"
Beta (β) Persei.	Algol	2.09–3.40	$03^h08.2^m$	+40° 57'	Eclipsing variable star

Deep-Sky Objects

Designation	Alternate name	Vis. mag	RA	Dec.	Description
Melotte 20	The Alpha Persei Stream	2.09–3.40	$03^h08.2^m$	+40° 57'	Open cluster
NGC 869/884	Double cluster	5.3/6.1	$02^h19.0^m/22.4^m$	+57° 09'/07'	Open cluster
NGC 1039	Messier 34	5.2	$02^h42.0^m$	+42° 47'	Open cluster
Trumpler 2		5.9	$02^h37.3^m$	+55° 29'	Open cluster
NGC 744		7.9	$01^h58.4^m$	+55° 29'	Open cluster
NGC 1220		11.8	$03^h11.7^m$	+53° 20'	Open cluster
King 5		–	$03^h14.5^m$	+52° 43'	Open cluster
NGC 1245	Herschel 25	8.4	$03^h14.7^m$	+47° 15'	Open cluster
NGC 1342	Herschel 88	6.7	$03^h31.6^m$	+37° 20'	Open cluster
NGC 1513	Herschel 60	8.4	$04^h10.0^m$	+49° 31'	Open cluster
NGC 1528	Herschel 61	6.4	$04^h15.4^m$	+51° 14'	Open cluster
NGC 1545		6.2	$04^h20.9^m$	+50° 15'	Open cluster
NGC 650–51	Messier 76	10.1	$01^h42.4^m$	+51° 34'	Planetary nebula

			RA	Dec.	Description
IC 351	PK 159–15.1	12.0	$03^h47.5^m$	+35° 03'	Planetary nebula
IC 2003	PK 161–14.1	12.5	$03^h56.4^m$	+33° 52'	Planetary nebula
NGC 1499	California Nebula	–	$04^h00.7^m$	+36° 37'	Emission nebula
NGC 1491	Herschel 258	–	$04^h03.4^m$	+51° 19'	Emission nebula
NGC 1333		–	$03^h29.3^m$	+31° 25'	Reflection nebula
NGC 1023	Herschel 156	9.3	$02^h30.4^m$	+39° 04'	Galaxy
NGC 1275	Perseus A	11.9	$03^h19.8^m$	+41° 31'	Galaxy

Objects in Auriga

Stars

Designation	Alternate name	Vis. mag	RA	Dec.	Description
Omega (ω) Aurigae.		5.01, 8.18	$04^h59.3^m$	+37° 53'	PA 359°; Sep. 5.4"
14 Aurigae		5.1, 7.4	$05^h15.4^m$	+32° 31'	PA 352°; Sep. 14.6"
UV Aurigae		7.7–10.6, 11.5	$05^h21.8^m$	+32° 31'	PA 313°; Sep. 3.6"
Theta (θ) Aurigae	37 Aurigae	2.6, 7.1	$05^h59.7^m$	+37° 13'	PA 313°; Sep. 3.6"
Psi⁵ (φ⁵) Aurigae		5.01, 8.18	$04^h59.3^m$	+37° 53'	PA 8°; Sep. 8.4"
Σ 928	Struve 928	7.6, 8.2	$06^h34.7^m$	+38° 32'	PA 133°; Sep. 3.5"
Σ 929	Struve 929	7.2, 8.3	$06^h35.4^m$	+37° 43'	PA 25°; Sep. 6.0"
Σ 698	Struve 698	6.6, 8.7	$05^h25.2^m$	+34° 51'	PA 345°; Sep. 31.2"
OΣ 147		5.6, 10. 10.6	$06^h34.3^m$	+38° 05'	PA 73°[AB], Sep. 43.2"[AB] / PA 117°[AC], Sep. 46.3"[AC]
AE Aurigae		5.4–6.1	$05^h16.3^m$	+34° 05'	Variable star
Beta (β) Aurigae		1.90	$05^h59.5^m$	+44° 57'	Variable star
RT Aurigae		5.0–5.8	$06^h28.6^m$	+30° 30'	Variable star
Alpha (α) Aurigae	Capella	0.08ᵥ	$05^h16.7^m$	+46° 00'	6th brightest star

Objects in Auriga (continued)

Deep-Sky Objects

Designation	Alternate name	Vis. mag	RA	Dec.	Description
NGC 1664	Herschel 59	7.6	$04^h51.1^m$	$+43°\ 42'$	Open cluster
NGC 1778	Herschel 61	7.7	$05^h08.1^m$	$+37°\ 03'$	Open cluster
NGC 1857	Herschel 33	7.0	$05^h20.2^m$	$+39°\ 21'$	Open cluster
NGC 1893		7.5	$05^h22.7^m$	$+33°\ 24'$	Open cluster
NGC 1883		12.0	$05^h25.9^m$	$+46°\ 33'$	Open cluster
NGC 1912	Messier 38	6.4	$05^h28.7^m$	$+35°\ 50'$	Open cluster
NGC 1907		8.2	$05^h28.0^m$	$+35°\ 19'$	Open cluster
NGC 1960	Messier 36	6.0	$05^h36.1^m$	$+34°\ 08'$	Open cluster
NGC 2099	Messier 37	5.6	$05^h24.4^m$	$+32°\ 33'$	Open cluster
NGC 2126	Herschel 68	10.2p	$06^h03.0^m$	$+49°\ 54'$	Open cluster
Harrington 4		–	05^h19^m	$+33°$	Open cluster
Stock 10		–	$05^h39.0^m$	$+37°\ 56'$	Open cluster
NGC 2281		5.4	$06^h49.3^m$	$+41°\ 04'$	Open cluster
Collinder 62		4.2p	$05^h22.5^m$	$+41°\ 00'$	Open cluster
IC 410		–	$05^h22.6^m$	$+34°\ 31'$	Emission nebula
IC 417		–	$05^h28.1^m$	$+34°\ 26'$	Emission nebula
NGC 1931	Herschel 261	–	$05^h31.4^m$	$+34°\ 15'$	Reflection & emission nebula
IC 405	Caldwell 31/Flaming Star Nebula	–	$05^h16.2^m$	$+34°\ 16'$	Reflection & emission nebula
SH2-224		–	$05^h27.3^m$	$+42°\ 59'$	SNR
IC 2149	PK166+10.1	10.7	$05^h56.3^m$	$+46°\ 07'$	Planetary nebula

Objects in Taurus

Stars

Designation	Alternate name	Vis. mag	RA	Dec.	Description
RW Tauri		7.9–11.4	04^{h}03.9^m	+28° 08'	Variable star
Phi (φ) Tauri		5.0, 8.4	04^{h}20.4^m	+27° 21'	PA 250°; Sep. 52.1″
Σ 559	Struve 559	7.0, 7.0	04^{h}33.5^m	+18° 01'	PA 277°; Sep. 3.2″
Alpha (α) Tauri	Aldebaran	0.87, 13.4	04^{h}35.9^m	+16° 31'	PA 34°; Sep. 121.7″
Σ 572	Struve 572	7.3, 7.3	04^{h}38.5^m	+26° 56'	PA 195°; Sep. 4.1″
118 Tauri		5.9, 6.7	05^{h}29.3^m	+25° 09'	PA 206°; Sep. 4.9″

Deep-Sky Objects

Designation	Alternate name	Vis. mag	RA	Dec.	Description
NGC 1514	Pk165–15.1	10.9p	04^{h}09.2^m	+30° 47'	Planetary nebula
Collinder 50	Melotte 25/Hyades	0.5	04^{h}27^m	+16°	Open cluster
NGC 1746		6.1	05^{h}03.6^m	+23° 49'	Open cluster
NGC 1807	Melotte 28	7.0	05^{h}10.7^m	+16° 32'	Open cluster
NGC 1817		7.7	05^{h}12.1^m	+16° 42'	Open cluster
Dolidze–Dzimselejsvili 3		–	05^{h}33.7^m	+26° 29'	Open cluster
Dolidze–Dzimselejsvili 4		–	05^{h}35.9^m	+25° 57'	Open cluster
NGC 1952	Messier 1/Crab Nebula	8.8	05^{h}03.6^m	+23° 49'	SNR

Objects in Gemini

Stars

Designation	Alternate name	Vis. mag	RA	Dec.	Description
BU Geminorum		5.7–7.5	$06^h12.3^m$	+22° 54'	Variable star
U Geminorum		8.2–14.9	$07^h55.1^m$	+22° 00'	Variable star
Zeta (ζ) Geminorum		3.6–4.1 0	$7^h04.1^m$	+20° 34'	Variable star
15 Geminorum		6.6, 8.0	$6^h27.8^m$	+20° 47'	PA 204°; Sep. 27.1"
20 Geminorum		6.3, 6.9	$6^h32.3^m$	+17° 47'	PA 210°; Sep. 20.0"
38 Geminorum		4.7, 7.7	$6^h54.6^m$	+13° 11'	PA 145°; Sep. 7.1"
Σ 1108	Struve 1108	6.6, 8.3	$7^h32.8^m$	+22° 53'	PA 178°; Sep. 11.5"
Delta (δ) Geminorum		3.6, 10.7	$7^h18.1^m$	+16° 32'	PA 33°; Sep. 9.6"

Deep-Sky Objects

Designation	Alternate name	Vis. mag	RA	Dec.	Description
NGC 2168	Messier 35	5.1	$06^h08.9^m$	+24° 20'	Open cluster
NGC 2158	Collinder 81	8.6	$06^h07.5^m$	+24° 06'	Open cluster
NGC 2129	Collinder 77	6.7	$06^h01.0^m$	+23° 18'	Open cluster
IC 2157	Collinder 80	8.4	$06^h05.0^m$	+24° 00'	Open cluster
NGC 2304	Herschel 2	10.0p	$06^h55.0^m$	+18° 01'	Open cluster
NGC 2266	Herschel 21	9.5p	$06^h43.2^m$	+26° 58'	Open cluster
NGC 2331	Collinder 126	8.5p	$07^h07.2^m$	+27° 21'	Open cluster
Collinder 89		5.7p	$06^h18.0^m$	+23° 38'	Open cluster
NGC 2355	Herschel 6	9.7p	$01^h16.9^m$	+13° 47'	Open cluster
NGC 2395	Collinder 144	8.0	$07^h27.1^m$	+13° 35'	Open cluster
NGC 2392	Caldwell 39/Eskimo Nebula	9.2	$07^h29.2^m$	+20° 55'	Planetary nebula

Objects in Orion

Stars

Designation	Alternate name	Vis. mag	RA	Dec.	Description
Rho (ρ) Orionis		4.5, 8.3	05h 13.3m	+02° 52'	PA 64°; Sep. 7.0"
Delta (δ) Orionis		2.0, 6.9	05h 32.0m	−00° 18'	PA 359°; Sep. 52.6"
Sigma (σ) Orionis		4.0, 10.3	05h 38.7m	−02° 36'	PA 238°; Sep. 11.4"
Zeta (ζ) Orionis		1.9, 4.0	05h 40.8m	−01° 57'	PA 165°; Sep. 2.3"
Iota (ι) Orionis		2.8, 7.3	05h 35.4m	−05° 55'	PA 141°; Sep. 11.3"
Eta (η) Orionis		3.6, 5.0	05h 24.5m	−02° 24'	PA 80°; Sep. 1.5"
Lambda (λ) Orionis	Dawes 5	3.5, 5.6	05h 35.1m	+09° 56'	PA 43°; Sep. 4.4"
Theta¹ (θ¹) Orionis	Trapezium	6.7, 7.9^{AB}/5.1, 6.7^{CD}	05h 35.3m	−05° 23'	PA 31°; Sep. 8.8"^{AB} / PA 241°, Sep. 13.4"^{CD}
Alpha (α) Orionis	Betelgeuse	0.40–1.3	05h 55.2m	+07° 24'	Variable star

Deep-Sky Objects

Designation	Alternate name	Vis. mag	RA	Dec.	Description
Collinder 70		0.4	05h 35.0m	−01° 56	Open cluster
Collinder 69		2.8p	05h 35.1m	+09° 56'	Open cluster
NGC 1981	Collinder 73	4.6	05h 35.2m	−04° 26'	Open cluster
NGC 2169	Collinder 38	5.9	06h 08.4m	+13° 57'	Open cluster
NGC 1976	Messier 42/Orion Nebula	2.9	05h 35.4m	−05° 27'	Emission & reflection nebula
NGC 1973 −75 −77		~4.7	05h 35.1m	−04° 44'	Emission & reflection nebula
NGC 1980		—	05h 35.4m	−05° 54'	Emission nebula
NGC 1982	Messier 43	6.9	05h 35.6m	−05° 16'	Emission & reflection nebula
NGC 1999		—	05h 36.5m	−06° 42'	Emission & reflection nebula
NGC 2024		—	05h 40.7m	+02° 27'	Emission nebula
NGC 2068	Messier 78	8	05h 46.7m	+00° 03'	Emission & reflection nebula
NGC 2023		—	05h 41.6m	−02° 16'	Emission & reflection nebula
IC 430		—	05h 38.5m	−07° 05'	Reflection nebula

Objects in Orion (*continued*)

Stars

Designation	Alternate name	Vis. mag	RA	Dec.	Description
–	$05^h\,38.5^m$	$-07°\,05'$	Reflection nebula		IC 430
IC 431		–	$05^h\,40.3^m$	$-01°\,27'$	Reflection nebula
IC 432		–	$05^h\,38.5^m$	$-07°\,05'$	Reflection nebula
Barnard 33	Horsehead Nebula	–	$05^h\,40.9^m$	$-02°\,28'$	Dark nebula
NGC 2022	PK 196–10.1	11.9	$05^h\,42.1^m$	$+09°\,05'$	Planetary nebula
Sharpless 2–276	Barnard's Loop	–	$05^h\,56.0^m$	$-02°\,00'$	SNR

Appendix 1
Astronomical Coordinate Systems

A basic requirement for studying the heavens is determining where in the sky things are. To specify sky positions, astronomers have developed several coordinate systems. Each uses a coordinate grid projected on to the celestial sphere, in analogy to the geographic coordinate system used on the surface of the Earth. The coordinate systems differ only in their choice of the fundamental plane, which divides the sky into two equal hemispheres along a great circle (the fundamental plane of the geographic system is the Earth's equator). Each coordinate system is named for its choice of fundamental plane.

The Equatorial Coordinate System

The equatorial coordinate system is probably the most widely used celestial coordinate system. It is also the one most closely related to the geographic coordinate system, because they use the same fundamental plane and the same poles. The projection of the Earth's equator onto the celestial sphere is called the celestial equator. Similarly, projecting the geographic poles on to the celestial sphere defines the north and south celestial poles.

However, there is an important difference between the equatorial and geographic coordinate systems: the geographic system is fixed to the Earth; it rotates as the Earth does. The equatorial system is fixed to the stars, so it appears to rotate across the sky with the stars, but of course it's really the Earth rotating under the fixed sky.

The latitudinal (latitude-like) angle of the equatorial system is called **declination** (Dec for short). It measures the angle of an object above or below the celestial equator. The longitudinal angle is called the **right ascension** (RA for short). It measures the angle of an object east of the vernal equinox. Unlike longitude, right ascension is usually measured in hours instead of degrees, because the apparent rotation of the equatorial coordinate system is closely related to sidereal time and hour angle. Since a full rotation of the sky takes 24 hours to complete, there are (360 degrees/24 hours) = 15 degrees in one hour of right ascension.

This coordinate system is illustrated in Figure A.1

The Galactic Coordinate System

The galactic coordinate system has latitude and longitude lines, similar to what you are familiar with on Earth. In the galactic coordinate system the Milky Way uses as its fundamental plane the zero

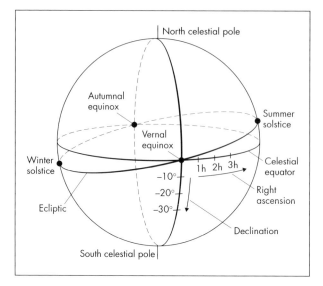

Figure A.1. The equatorial coordinate system.

degree latitude line in the plane of our Galaxy, and the zero degree longitude line is in the direction of the center of our galaxy. The latitudinal angle is called the galactic latitude, and the longitudinal angle is called the galactic longitude. This coordinate system is useful for studying the Galaxy itself.

The reference plane of the galactic coordinate system is the disk of our Galaxy (i.e. the Milky Way) and the intersection of this plane with the celestial sphere is known as the galactic equator, which is inclined by about 63° to the celestial equator. **Galactic latitude,** b (see Figure A.2) is analogous to declination, but measures distance north or south of the galactic equator, attaining +90° at the north galactic pole (NGP) and –90° at the south galactic pole (SGP).

Galactic longitude, l, is analogous to right ascension and is measured along the galactic equator in the same direction as right ascension. The zero point of galactic longitude is in the direction of the galactic center (GC), in the constellation of Sagittarius; it is defined precisely by taking the galactic longitude of the north celestial pole to be exactly 123°. This somewhat confusing system is best shown by the diagram in Figure A.2.

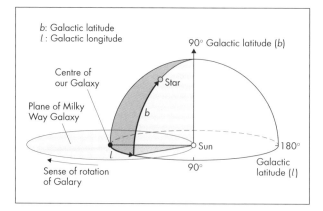

Figure A.2. The Galactic coordinate system.

Appendix 2

Magnitudes

The first thing that strikes even a casual observer is that the stars are of differing brightness. Some are faint, some are bright, and a few are very bright; this brightness is called the magnitude of a star.

The origins of this brightness system are historical, when all the stars seen with the naked eye were classified into one of six magnitudes, with the brightest being called a "star of the first magnitude", the faintest a "star of the sixth magnitude". Since then the magnitude scale has been extended to include negative numbers for the brightest stars, and decimal numbers used between magnitudes, along with a more precise measurement of the visual brightness of the stars. Sirius has a magnitude of –1.44, while Regulus has a magnitude of +1.36. Magnitude is usually abbreviated to m. Note that the brighter the star, the smaller the numerical value of its magnitude.

A difference between two objects of 1 magnitude means that the object is about 2.512 times brighter (or fainter) than the other. Thus a first-magnitude object (magnitude m = 1) is 2.512 times brighter than a second-magnitude object (m = 2). This definition means that a first-magnitude star is brighter than a sixth-magnitude star by the factor 2.512 raised to the power of 5. That is a hundredfold difference in brightness. The naked-eye limit of what you can see is about magnitude 6, in urban or suburban skies. Good observers report seeing stars as faint as magnitude 8 under exceptional conditions and locations. The magnitude brightness scale doesn't tell us whether a star is bright because it is close to us, or whether a star is faint because it's small or because it's distant. All that this classification tells us is the apparent magnitude of the object – that is, the brightness of an object as observed visually, with the naked eye or with a telescope. A more precise definition is the absolute magnitude, M, of an object. This is defined as the brightness an object would have at a distance of 10 parsecs from us. It's an arbitrary distance, deriving from the technique of distance determination known as parallax; nevertheless, it does quantify the brightness of objects in a more rigorous way. For example, Rigel has an absolute magnitude of –6.7, and one of the faintest stars known, Van Biesbroeck's Star, has a value of +18.6.

Of course, the above all assumes that we are looking at objects in the visible part of the spectrum. It shouldn't come as any surprise to know that there are several further definitions of magnitude that rely on the brightness of an object when observed at a different wavelength, or waveband, the most common being the U, B and V wavebands, corresponding to the wavelengths 350, 410 and 550 nanometers respectively. There is also a magnitude system based on photographic plates: the photographic magnitude, m_{pg}, and the photovisual magnitude, m_{pv}. Finally, there is the bolometric magnitude, $m_{BOL,}$ which is the measure of all the radiation emitted from the object.[1]

---

[1] It is interesting to reflect that *no* magnitudes are in fact a true representation of the brightness of an object, because every object will be dimmed by the presence of interstellar dust. All magnitude determinations therefore have to be corrected for the presence of dust lying between us and the object. It is dust that stops us from observing the center of our Galaxy.

Objects such as nebulae and galaxies are **extended objects**, which means that they cover an appreciable part of the sky: in some cases a few degrees, in others only a few arc minutes. The light from, say, a galaxy is therefore "spread out" and thus the quoted magnitude will be the magnitude of the galaxy were it the "size" of a star; this magnitude is often termed the **combined** or **integrated magnitude**. This can cause confusion, as a nebula with, say, a magnitude of 8, will appear fainter than an 8th magnitude star, and in some cases, where possible, the surface brightness of an object will be given. This will give a better idea of what the overall magnitude of the object will be.

Finally, many popular astronomy books will tell you that the faintest, or limiting magnitude, for the naked eye is around the 6th magnitude. This may well be true for those of us who live in an urban location. But the truth of the matter is that from exceptionally dark sites with a complete absence of light pollution, magnitudes as faint as 8 can be seen. This will come as a surprise to many amateurs. Furthermore, when eyes are fully dark-adapted, the technique of averted vision will allow you to see with the naked eye up to three magnitudes fainter! But before you rush outside to test these claims, remember that to see really faint objects, either with the naked eye or telescopically, several other factors such as the transparency and seeing conditions, and the psychological condition of the observer (!) will need to be taken into consideration, with light pollution as the biggest evil.

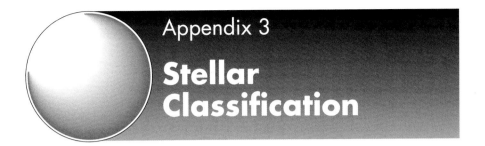

Appendix 3
Stellar Classification

For historical reasons a star's classification is designated by a capital letter thus, in order of decreasing temperature:

O B A F G K M L R N S

The sequence goes from hot blue stars types O and A to cool red stars K and M. and L. In addition, there are rare and hot stars called Wolf–Rayet stars, class WC and WN, exploding stars Q, and peculiar stars, P. The star types R, N and S actually overlap class M, and so R and N have been reclassified as C-type stars, the C standing for carbon stars. A new class has recently been introduced, the L class.[2] Furthermore, the spectral types themselves are divided into ten spectral classes beginning with 0, 1, 2, 3 and so on up to 9. A class A1 star is thus hotter than a class A8 star, which in turn is hotter than a class F0 star. Further prefixes and suffixes can be used to illustrate additional features:

a star with:

emission lines	e (also called f in some O-type stars)
metallic lines	m
a peculiar spectrum	p
a variable spectrum	v
a blue or shift in the line (for example, P-Cygni stars)	q

And so forth. For historical reasons, the spectra of the hotter star types O, A and B are sometimes referred to as **early-type** stars, while the cooler ones, K, M, L, C and S, are later-type. Also, F and G stars are **intermediate-type** stars.

---

[2] These are stars with very low temperatures: 1900–1500 K. Many astronomers believe these are brown dwarves.

Appendix 4

Light Filters

One of the most useful accessories an amateur can possess is one of the ubiquitous optical filters. Having been accessible previously only to the professional astronomer, they came on to the market relatively recently, and have made a very big impact. They are useful, but don't think they're the whole answer! They can be a mixed blessing. From reading some of the advertisements in astronomy magazines you would be correct in thinking that they will make hitherto faint and indistinct objects burst into vivid observability –sometimes!

What the manufacturers do not mention is that regardless of the filter used, you will still need dark and transparent skies for the use of the filter to be worthwhile. Don't make the mistake of thinking that using a filter from an urban location will always make objects become clearer. The first and most immediately apparent item on the downside is that in all cases the use of a filter reduces the total amount of light that reaches the eye, often quite substantially. However, what the filter does do is select specific wavelengths of light emitted by an object, which may be swamped by other wavelengths. It does this by suppressing the unwanted wavelengths. This is particularly effective when observing extended objects such as emission nebulae and planetary nebulae.

In the former case, use a filter that transmits light around the wavelength of 653.2 nm, which is the spectral line of hydrogen alpha (Hα), and is the wavelength of light responsible for the spectacular red color seen in photographs of emission nebulae. Some filters may transmit light through perhaps two wavebands: 486 nm for hydrogen beta[3] (Hβ) and 500.7 nm for oxygen-3 [OIII], two spectral lines which are very characteristic in planetary nebula. Use of such filters will enhance the faint and delicate structure within nebulae, and, from a dark site, they really do bring out previously invisible detail. Don't forget (as the advertisers sometimes seem to) that "nebula" filters do not (usually) transmit the light from stars, and so when in use the background will be dark with only nebulosity visible, and this makes them somewhat redundant for observing stars, star clusters and galaxies alone, unless the aforementioned objects are associated with nebulosity, as can often be the case.

One kind of filter that does help in heavily light-polluted areas is the LPR (light-pollution reduction filter), which effectively blocks out the light emitted from sodium and mercury street lamps, at wavelengths 366, 404.6, 435.8, 546.1, 589.0 and 589.5 nm. Clearly the filter will only be effective if the light from the object you want to see is significantly different from the light-polluting source: fortunately, this is usually the case. Light-pollution reduction filters can be very effective visually and photographically, but remember that there is always some overall reduction in brightness of the object you are observing.

---

[3]This filter can be used to view dark nebulae that are overwhelmed by the proximity of emission nebulae. A case in point is the Horsehead Nebula, which is incredibly faint, and swamped by light from the surrounded emission nebulosity.

Whatever filters you decide on, it is worthwhile trying to use them before you make a purchase (they are expensive!), by borrowing them either from a fellow amateur or from a local astronomical society. This will show you whether the filter really makes any difference to your observing.

There is no doubt that modern filters can be an excellent purchase, but it may be that your location or other factors will prevent the filter from realizing its full potential or value for money. Most commercially available filters are made for use at a telescope and not for binoculars, so unless you are mechanically minded and can make your own filter mounts (and are happy to pay – two LPR filters could easily cost more than the binoculars!), it's likely that only those observers with telescopes can benefit.

Appendix 5

Star Clusters

Open or **galactic clusters**, as they are sometimes called, are collections of young stars, containing anything from maybe a dozen members to hundreds. A few of them, for example, Messier 11 in Scutum, contain an impressive number of stars, equaling that of globular clusters (see below), while others seem little more than a faint grouping set against the background star field. Such is the variety of open clusters that they come in all shapes and sizes. Several are over a degree in size and their full impact can only be appreciated by using binoculars, as a telescope has too narrow a field of view. An example of such a large cluster is Messier 44 in Cancer. Then there are tiny clusters, seemingly nothing more than compact multiple stars, as is the case with IC 4996 in Cygnus. In some cases all the members of the cluster are equally bright, such as Caldwell 71 in Puppis, but there are others that consist of only a few bright members accompanied by several fainter companions, as in the case of Messier 29 in Cygnus. The stars which make up an open cluster are called **Population I** stars, which are metal-rich[4] and usually to be found in or near the spiral arms of the Galaxy.

The reason for the varied and disparate appearances of open clusters is the circumstances of their births. It is the interstellar material out of which stars form that determines both the number and type of stars that are born within it. Factors such as the size, density, turbulence, temperature, and magnetic field all play a role as the deciding parameters in star birth. In the case of **giant molecular clouds**, or GMCs, the conditions can give rise to both O- and B-type giant stars along with solar-type dwarf stars – whereas in **small molecular clouds** (SMCs) only solar-type stars will be formed, with none of the luminous B-type stars. An example of an SMC is the **Taurus Dark Cloud**, which lies just beyond the Pleiades.

An interesting aspect of open clusters is their distribution in the night sky. Surveys show that although well over a thousand clusters have been discovered, only a few are observed to be at distances greater than 25° above or below the galactic equator. Some parts of the sky are very rich in clusters – Cassiopeia and Puppis – and this is due to the absence of dust lying along these lines of sight, allowing us to see across the spiral plane of our Galaxy. Many of the clusters mentioned here actually lie in different spiral arms, and so as you observe them you are actually looking at different parts of the spiral structure of our Galaxy.

An open cluster presents a perfect opportunity for observing star colors (see Appendix 7). Many clusters, such as the ever and rightly popular Pleiades, are all a lovely steely blue color. On the other hand, Caldwell 10 in Cassiopeia has contrasting bluish stars along with a nice orange star. Other clusters have a solitary yellowish or ruddy orange star along with fainter white ones, such as Messier 6 in Scorpius. An often striking characteristic of open clusters is the apparent chains of stars that are seen. Many clusters have stars that arc across apparently empty voids, as in Messier 41 in Canis Major. Another word for a very small, loose group of stars is an **asterism**. In some cases there may only be five or six stars within the group.

---

[4] Astronomers call every element other than hydrogen and helium, metals.

Open clusters are groups of stars that are usually young and have an appreciable angular size and may have a few hundred components. **Globular clusters** are clusters that are very old, are compact and may contain up to a million stars, and in some cases even more. The stars that make up a globular cluster are called **Population II** stars. These are metal-poor stars and are usually to be found in a spherical distribution around the galactic center at a radius of about 200 light years. Furthermore, the number of globular clusters increases significantly the closer one gets to the galactic center. This means that particular constellations that are located in a direction towards the galactic bulge have a high concentration of globular clusters within them, such as Sagittarius and Scorpius.

The origin and evolution of a globular cluster are very different from an open or galactic cluster. All the stars in a globular cluster are very old, with the result that any star earlier than a G or F type star will have already left the main sequence and be moving toward the red giant stage of its life. In fact, new star formation no longer takes place within any globular clusters in our Galaxy, and they are believed to be the oldest structures in our Galaxy. In fact, the youngest of the globular clusters is still far older than the oldest open cluster. The origin of the globular clusters is a topic of fierce debate and research, with the current models predicting that the globular clusters may have been formed within the protogalaxy clouds that went to make up our Galaxy.

There are about 150 globular clusters ranging in size from 60 to 150 light years in diameter. They all lie at vast distances from the Sun, and are about 60,000 light years from the galactic plane. The nearest globular clusters, for example Caldwell 86 in Ara, lie at a distance of over 6000 light years, and thus the clusters are difficult objects for small telescopes.

Appendix 6

Double Stars

Double stars are stars that although they appear to be just one single star, will on observation with either binoculars or telescopes resolve themselves into two stars. Many stars may appear as double due to them lying in the same line of sight as seen from the Earth, and this can only be determined by measuring the spectra of the stars and calculating their red (or blue) shifts. Such stars are called **optical doubles**. It may well be that the two stars are separated in space by a vast distance. Some, however, are actually gravitationally bound and may orbit around each other, over a period of days or even years.

Many double stars cannot be resolved by even the largest telescopes, and are called **spectroscopic binaries**, the double component only being fully understood when the spectra are analyzed. Others are **eclipsing binaries**, such as Algol (β Persei), where one star moves during its orbit in front of its companion, thus brightening and dimming the light observed. A third type is the **astrometric binary,** such as Sirius (α Canis Majoris), where the companion star may only be detected by its influence on the motion of the primary star.

The brighter of the two stars is usually called the **primary** star, whilst the fainter is called the **secondary** or **companion**. This terminology is employed regardless of how massive either star is, or whether the brighter is in fact the less luminous of the two in reality, but just appears brighter as it may be closer.

Perhaps the most important terms used in double-star work are the **separation** and the **position angle** (PA). The separation is the angular distance between the two stars, usually in seconds of arc, and measured from the brighter star to the fainter. The position angle is the relative position of one star, usually the secondary, with respect to the primary, and is measured in degrees, with 0° at due north, 90° at due east, 180° due south, 270° at due west, and back to 0°. It is best described by an example (see Figure A.3), the double star γ Virginis, with components of magnitude 3.5 and 3.5, has a separation of 1.8 arcseconds, at a PA of 267° (epoch 2000.0). Note that the secondary star is the one always placed somewhere on the orbit, with the primary star at the center of the perpendicular lines. The separation and PA of any double star are constantly changing, and should be quoted for the year observed. When the period is very long, some stars will have no appreciable change in PA for several years; others, however, will change from year to year.

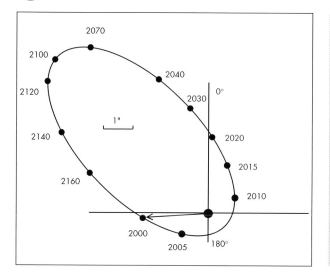

Figure A.3. The motion of γ Virginis.

Appendix 7

Star Colors

The most important factor which determines what the color of a star is, is you – the observer! It is purely a matter of both physiological and psychological influences. What one observer describes as a blue star, another may describe as a white star, or one may see an orange star, whilst another observes the same star as being yellow. It may even be that you will observe a star to have different color when using different telescopes or magnifications, and atmospheric conditions will certainly have a role to play. The important thing to remember is that whatever color you observe a star to have, then that is the color you should record.[5]

It may seem to a casual observer that the stars do not possess many bright colors, and only the brightest stars show any perceptible color: Betelgeuse can be seen to be red, and Capella is yellow, whilst Vega is blue, and Aldebaran has an orange tint, but beyond that most stars seem to be an overall white. To the naked eye, this is certainly the case, and it is only with some kind of optical equipment that the full range of star color becomes apparent.

But what is meant by the color of a star? A scientific description of a star's color is one that is based on the stellar classification, which in turn is dependent upon the chemical composition and temperature of a star. In addition, a term commonly used by astronomers is the color index. This is determined by observing a star through two filters, the B and the V filters, which correspond to wavelengths of 440 nm and 550 nm respectively, and measuring its brightness. Subtracting the two values obtained gives B – V, the **color index**. Usually, a blue star will have a color index that is negative, i.e. –0.3, orange-red stars could have a value greater than 0.0, and upwards to about 3.00 and greater for very red stars (M6 and greater). But as this is an observationally based book, the scientific description will not generally apply.

As mentioned above, red, yellow, orange and blue stars are fairly common, but are there stars which have, say, a purple tint, or blue, or violet, crimson, lemon, and the ever elusive green color? The answer is yes, but with the caution that it depends on how you describe the color. A glance at the astronomy books from the nineteenth century and beginning of the twentieth century will show you that star color was a hot topic, and descriptions such as amethyst (purple), cinerous (wood-ash tint), jacinth (pellucid orange), and smalt (deep blue), to name but a few, were used frequently. Indeed, the British Astronomical Association even had a section devoted to star colors. But today, observing and cataloging star color is just a pleasant pastime. Nevertheless, under good seeing conditions, with a dark sky, the keen-eyed observer will be able to see the faint tinted colors from deepest red to steely blue, with all the colors in between.

It is worth noting that several distinctly colored stars occur as part of a double-star system. The reason for this may be that although the color is difficult to see in an individual star, it may appear

---

[5] An interesting experiment is to observe a colored star first through one eye and then the other. You may be surprised by the result!

more intense when seen together with a contrasting color. Thus, in the section on double and triple stars, there are descriptions of many beautifully colored systems. For instance, the fainter of the two stars in η Cassiopeiae has a distinct purple tint, whilst in γ Andromadae and α Herculis, the fainter stars are most definitely green.

Books, Magazines and Astronomical Organizations

I have selected a few of the many fine astronomy and astrophysics books in print that I believe are amongst the best on offer. I do not expect you to buy, or even read them all, but check at your local library to see if they have some of them.

Star Atlases and Observing Guides

Amateur Astronomer's Handbook, J. Sidgwick, Pelham Books, London, UK, 1979
Burnham's Celestial Handbook, R. Burnham, Dover Books, New York, USA, 1978
Deep-Sky Companions: The Messier Objects, S. O'Meara, Cambridge University Press, Cambridge, UK, 1999
Millennium Star Atlas, R. Sinnott, M. Perryman, Sky Publishing, Massachusetts, USA, 1999
Norton's Star Atlas & Reference Handbook, I. Ridpath (Ed.), Longmans, Harlow, UK, 1999
Observing Handbook and Catalogue of Deep-Sky Objects, C. Luginbuhl, B. Skiff, Cambridge University Press, Cambridge, USA, 1990
Observing the Caldwell Objects, D. Ratledge, Springer-Verlag, London, UK, 2000
Sky Atlas 2000.0, W. Tirion, R. Sinnott, Sky Publishing & Cambridge University Press, Massachusetts, USA, 1999
The Night Sky Observer's Guide, Vols. I and II, G. Kepple, G. Sanner, Willman-Bell, Richmond, USA, 1999
Uranometria 2000.0 Volumes 1 & 2, Wil Tirion (Ed), Willmann-Bell, Virginia, USA, 2001

Astronomy and Astrophysics Books

Field Guide to the Deep Sky Objects, M. D. Inglis, Springer, London, UK, 2001
Galaxies and Galactic Structure, D. Elmegreen, Prentice Hall, New Jersey, USA, 1998
Introductory Astronomy & Astrophysics, M. Zeilik, S. Gregory, E. Smith, Saunders College Publishing, Philadelphia, USA, 1999

Observer's Guide to Stellar evolution, M. D. Inglis, Springer, London, UK, 2002
Stars, J. B. Kaler, Scientific American Library, New York, USA, 1998
Stars, Nebulae and the Interstellar Medium, C. Kitchin, Adam Hilger, Bristol, UK, 1987
The Milky Way, Bart & Priscilla Bok, Harvard Science Books, Massachusetts, USA, 1981
Voyages Through The Universe, A. Fraknoi, D. Morrison, S. Wolff, Saunders College Publishing, Philadelphia, USA, 2000

Magazines

Astronomy Now UK
　Sky & Telescope USA
　New Scientist UK
　Scientific American USA
　Science USA
　Nature UK

The first three magazines are aimed at a general audience and so are applicable to everyone; the last three are aimed at the well-informed lay person. In addition there are many research-level journals that can be found in university libraries and observatories.

Organizations

The Federation of Astronomical Societies, 10 Glan y Llyn, North Cornelly, Bridgend County Borough, CF33 4EF, Wales, UK
http://www.fedastro.demon.co.uk/

Society for Popular Astronomy, The SPA Secretary, 36 Fairway, Keyworth, Nottingham NG12 5DU, UK
http://www.popastro.com/

The American Association of Amateur Astronomers, P.O. Box 7981, Dallas, TX 75209-0981, USA
http://www.corvus.com

The Astronomical League
http://www.astroleague.org/

The British Astronomical Association, Burlington House, Piccadilly, London, W1V 9AG, UK.
http://www.ast.cam.ac.uk/~baa/

The Royal Astronomical Society, Burlington House, Piccadilly, London W1V 0NL, UK
http://www.ras.org.uk/membership.htm

The Webb Society
http://www.webbsociety.freeserve.co.uk/

International Dark-Sky Association, 3225 N. First Ave., Tucson, AZ 85719, USA.
http://www.darksky.org/

Campaign for Dark Skies, 38 The Vineries, Colehill, Wimborne, Dorset, BH21 2PX, UK.
http://www.dark-skies.freeserve.co.uk/

Appendix 9

The Greek Alphabet

The following is a quick reference guide to the Greek letters, used in the Bayer classification system. Each entry shows the uppercase letter, the lowercase letter, and the pronunciation.

A α	Alpha	H η	Eta	N ν	Nu	T τ	Tau
B β	Beta	Θ θ	Theta	Ξ ξ	Xi	Y υ	Upsilon
Γ γ	Gamma	I ι	Iota	O o	Omicron	Φ φ	Phi
Δ δ	Delta	K κ	Kappa	Π π	Pi	X χ	Chi
E ε	Epsilon	Λ λ	Lambda	P ρ	Rho	Ψ ψ	Psi
Z ζ	Zeta	M μ	Mu	Σ σ	Sigma	Ω ω	Omega

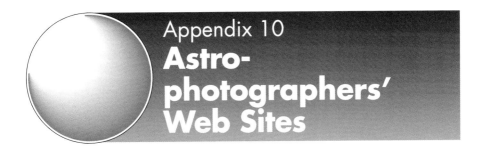

Appendix 10
Astro-photographers' Web Sites

Matt BenDaniel – http://starmatt.com/
Mario Cogo – www.intersoft.it/galaxlux
Bert Katzung – www.astronomy-images.com
Dr. Jens Lüdeman – http://www.ias–observatory.org/IAS/index–english.htm
Axel Mellinger – http://home.arcor–online.de/axel.mellinger/
Thor Olson – http://home.att.net/~nightscapes/photos/MilkyWayPanoramas/
Harald Straus [Astronomischer Arbeitskreis Salzkammergut] – http://www.astronomie.at/
Chuck Vaughn – http://www.aa6g.org/Astronomy/astrophotos.html
SEDS – http://www.seds.org/~spider/ngc/ngc.html

Index of Objects

The entry refers to its most familiar name and/or its main entry in the books. Page numbers are referred to by book number followed by page number.

CL